AI EVERYWHERE

Volume 1

How Women Are Changing the World with Artificial Intelligence

Six Peas Press

AI EVERYWHERE, VOLUME 1

Published by Six Peas Press, 2026, North Carolina, USA.

Disclaimer: This book and any related supplemental materials are not intended as a substitute for legal or professional advice. The content is for informational purposes only. The Publisher and Author make no representations or warranties with respect to the accuracy or completeness of the contents of this work and specifically disclaim all warranties, expressed or implied, including without limitation warranties of fitness for a particular purpose. This work is sold with the understanding that the Publisher and Author are not engaged in rendering legal, accounting, or other professional services. If professional assistance is required, the services of a competent professional person should be sought. Neither the Publisher nor the Author shall be liable for damages arising herefrom.

Mention of an organization, website, or person in this work as a reference or source of further information does not mean that the Author or the Publisher endorse the information or recommendations they may offer. Further, readers should be aware that websites listed in this work may have changed or disappeared between when this work was written and when it was read.

ISBN: 979-8-9996096-4-9 (ebook)

ISBN: 979-8-9996096-5-6 (paperback)

ISBN: 979-8-9996096-6-3 (hardcover)

Six Peas Press (SixPeas.net) is an imprint of **She Writes AI, LLC.** Email: books@SixPeas.net.

Website for the *She Writes AI Everywhere* book series: aiEverywhereBooks.com

Contents

BUSINESS

HOW-TO

AI EVERYWHERE
Volume 1

How Women Are Changing the World
with Artificial Intelligence

Introduction

She Writes Back

IN DECEMBER 2024, I saw a post on Substack highlighting 10 writers on AI to follow: 10 men, zero women. In a private message, the post writer told me he didn't know of any women writing about AI on Substack! So I published a list of 34 women writing about AI and data who I was already subscribed to, and sent him the link.

That list was the seed for *She Writes AI*. It began growing as people recommended other women writers. As of February 2026, the directory has 600+ women and nonbinary people affiliated with 60+ countries. The newsletter now shares weekly *She Writes AI* digests to subscribers (usually summarizing 200+ articles). Our volunteer organizers and allies actively seek ways to support and elevate the writing and initiatives of She Writes AI Community members (SheWritesAI, 2026).

Why was this book written?

In the AI arena, women and other demographic groups (as well as people outside the USA) are typically under-represented or mis-represented in three ways:

- In the data used for training AI,
- On the teams who develop AI platforms and solutions,
- Among the outputs generated by AI tools.

AI has great potential to be more objective than humans, to reduce biases and imbalances, and to bring widespread benefits to the world. However, as a result of the current poor representation, the dominant AI tools (especially generative AI) are more likely to reinforce and worsen existing biases, and to exacerbate inequities.

For more insights on these pros and cons, see my book, *Everyday Ethical AI: A Guide For Families & Small Businesses* (Smiley, 2025).

Women's presence and active participation in AI tend to bring out questions of fairness, care, and trust that profit-driven systems are designed to ignore. Across disciplines, women have consistently challenged the field to consider not just what technology can do, but what it should do and who it should serve.

In parallel, conflicting stories have surfaced in the past few years on whether women are under-using AI or have surpassed men's usage in some areas (Grau, 2025). An analysis of interview data from my "AI, Software, & Wetware" series showed no differences in AI usage by gender (Fischer, 2025; Smiley, 2026). Some studies report that women are penalized more severely for using AI than men are, motivating them to either use it less often or under-report when they do use it (Travis, 2025).

These factors are reflected in the dominance of the voices of Euro-American white men on AI topics. AI conference panels and speaker lineups often lack diversity, with organizers claiming they cannot find suitable women for those opportunities (OR Briefings, 2025; Schmidt, 2025; Murphy, 2026). Well, women experts in AI **are** out there. And they (we) are sharing their expertise with you here. This book is the first AI guide written exclusively by **women and nonbinary** authors.

In short, the world needs global women's voices and distinct perspectives on AI across all areas of daily life, and this book delivers them.

> **"The solution to biased AI is not to abandon the technology but to flood it with women's perspectives, experiences, and values." (Grau, 2025)**

Why is the book named "AI Everywhere"?

Our book is titled *AI Everywhere, Volume 1: How Women Are Changing The World With Artificial Intelligence* because AI is everywhere in our daily home and work lives, whether we realize it or not.

And women everywhere are using AI in creative ways to change the world–and writing about it. Our authors share a broad view of the many ways AI is impacting our home lives, our professional work, and our society. These chapters cover "What about AI in <topic>?" from the perspective of women and nonbinary writers with deep expertise in applying AI tools and technologies for those topics.

Identifying this volume as Volume 1 reflects the rapid pace of change in the world of AI. It is the first in a planned series of books by the She Writes AI Community on how women are applying AI to change the world.

Who is this book for?

This book is for working people of all ages and genders *(including parents and caregivers, paid or not; that's work too)*. No computer science degrees are needed to benefit from the book. If you want to better understand how AI is impacting our lives, and discover hidden AI biases and opportunities in everyday life and work, you'll find useful insights here. To the best of our knowledge, no other guide covers this breadth of applications of AI.

What's inside?

This book shares perspectives from 26 SheWritesAI Community members from five continents (13 of 26 authors are in countries outside the USA). Our experts share how they are using AI – and when & why they are not using AI–in a wide range of domains. If you've ever wondered how AI is really being used in a domain nowadays, check our chapter list for that topic and dive in! We cover:

AI in Society:

- **Celeste Garcia** recaps the history of how women have advanced AI technologies, and calls out what's needed for AI to be representative and accountable.
- **Rebecca Mbaya** highlights the alignments of Ubuntu, Yoruba, and Akan philosophies with AI frameworks and practices in the global south, based on the interviews she has conducted in The African Innovators Series.
- **Sarah Ennett** explains why that famous Venn diagram doesn't reflect what ikigai is, three scenes showing where AI is creating

ikigai risks for humans, and five things we can do.

- **Sydnor Hain** covers why folks in the west have become accustomed to "'Wonder Bread" ethics and how we can use AI with relational practices to replace mindless rules with more-nourishing ethics.
- **Elettra Fiumi** shows how AI empowers cinema artists to explore possibilities not previously open to them, with examples from her 10-year Radical Landscapes documentary as well as her award-winning AI film Mamma Robot.
- **Karen Spinner** covers why AI detection tools fail, how our brains know the difference between AI-written and human-written words, and why we need to ask different questions now about writing.

AI for Learning:

- **Cassandra Jens** and **Whitney Whealdon** share how they have used AI tools to help with K-12 homeschooling challenges, and how using these tools has affected their kids' skills and learning.
- **Jennifer W. Shewmaker, PhD** shares her experiences as a university provost with adopting AI in higher education.
- **Cheyenne Dominguez** explains why upskilling for women matters and how women can thrive by seizing current opportunities.

AI for Home and Health:

- **Andrea Hiott** identifies opportunities for AI chatbots to help us care more for others and find more caring connections beyond technology.
- **Midhat Tilawat** highlights how AI tools have provided decision support to her in two stressful areas of everyday life: postpartum and caregiving for an ailing parent.
- **Brie-Anna Willey** zooms in on burnout and three ways AI and algorithms shape our mental health and professional identities: attention, connection, and work.
- **Dhani Ramadhani** shares key parenting challenges in our AI-filled world and how she is raising humans who recognize their own agency and understand that AI is a tool, not an oracle.

AI in Business:

- An anonymous author shares their experiences with solopreneurship and how they have leveraged "AI Dream Teams" to solve practical issues faced by many entrepreneurs.
- **Elizabeth Eagle-Simbeye** covers AI and data privacy in automotive industries from the perspective of vehicle users (most of us) as well as vehicle builders.
- **Farida Khalaf** identifies how intelligent farming systems can leverage AI to help to feed the world.
- **Lisa Raehsler** explains how those uncannily-accurate ads keep popping up, how our data is being used in advertising & marketing, and what we can do to protect our privacy.
- **Katrina Watson** relates a brief history of how some retailers have been using AI to change how we shop, and shares why she thinks that new AI tools could boost smaller retailers.

AI How-To's:

These six chapters provide the most benefit to readers with a base level of technical expertise on the topic. The intended audience is called out at the start of each chapter.

- **Farida Khala**f highlights how AI tools impact both the threats and the solutions in cybersecurity.
- **Soribel Feliz** identifies how people with expertise in data privacy and cybersecurity are well-positioned to lead in AI governance.
- **Blessing Okpala, PhD** covers empathy and user experience, showing why AI tools must create inclusive outcomes and AI tools themselves must be accessible so that all people can actually use them.
- **Elena Calvillo** shares her hard-won lessons learned on how to make rapid AI prototyping safe and maintainable.
- **Lakshmi Veeramani** explains the technical aspects of agentic AI systems and what makes them such powerful tools, if well-managed.

- **Iwette Rapoport** spells out her methods for human-AI collaboration through governance in the chat threads with large language models.

All of our contributing authors are She Writes AI Community members who publish newsletters sharing their expertise on using AI and data in their fields.

Bottom line: This book not only sheds light on a broad range of practical uses of AI, it helps to dispel myths that "women don't do AI". The first chapter spotlights how women have contributed for decades (and still do) to the development of advanced AI technologies. The remaining chapters highlight ways the authors are using and advancing AI applications today, sharing lessons from their experiences that people of all genders can leverage in their home and work lives. Enjoy!

– Karen Smiley, founder and leader, She Writes AI (*shewritesai.org*)

Thank you to Hollie C. White for her skillful guidance on improving this Introduction and many of the chapters in this book. All remaining areas for improvement are the responsibility of the authors, including me.

About the Author

***Karen Smiley** is an AI leader, innovator, ethicist, data strategist, and author. During her 30+ year engineering career in data and software, she earned 14 granted US patents. Karen champions everyday ethical AI and inclusion in her AI newsletters. Her **AI, Software, & Wetware** podcast has drawn listeners in 100+ countries. Karen's debut book, **Everyday Ethical AI: A Guide For Families & Small Businesses**, was published in September 2025. She launched and leads the **She Writes AI** community of 600+ writers and initiated the She Writes AI Everywhere book series. She is based in North Carolina, USA.*

*To learn more about Karen Smiley and her work and writing, visit **karensmiley.com** or **everydayethicalai.com**, or find her at **substack.com/@karensmiley**. To learn more about the She Writes AI Community, visit **shewritesai.org** or **substack.com/@shewritesai**.*

SOCIETY

Chapter 1

The Hidden Architects

Challenging the Male Narrative and Building Accountable AI

By Celeste Garcia

THIS CHAPTER CHALLENGES the male-dominated narrative of AI's evolution by spotlighting the women who shaped the field from its earliest conceptual foundations to the researchers, engineers, and ethicists now steering AI through one of the most consequential technological shifts in history. As the AI community confronts challenges such as bias, labor exploitation, and inequality, women continue to lead the push for human-centered systems grounded in accountability and ethics. This work is not optional; it's imperative. Because AI systems are not neutral and reflect the biases, perspectives, and values of their creators, a lack of diversity in the development phase risks deepening gender and racial disparities and hardwiring structural inequities.

Pioneers of Universal Computation and Code

The true history of computing began with the realization that machines could manipulate abstract symbols—not just numbers—through algorithmic processes. This insight laid the foundation for modern computing and remains fundamental to today's AI systems, including large language models and generative AI. This concept can be traced to a woman who was a true visionary.

The concept that computing machines could move beyond simple calculation to perform any abstract operation—the defining phenomenon of computer technology—dates to the 1840s, thanks to **Ada Lovelace**, a pioneering mathematician and computer scientist.

Working with Charles Babbage on the theoretical Analytical Engine in the mid-19th century, Lovelace created what is considered the world's first computer algorithm. She surmised that a machine could manipulate symbols, not just numbers, allowing for an algebraic rather than just an arithmetic output. This profound realization—that numbers could represent entities other than quantity—launched the idea of universal computation, a century before it was realized. Even more prescient, Lovelace articulated the limitations, arguing that *"The Analytical Engine has no pretensions to originate anything. It can do whatever we know how to order it to perform."* Known as "Lady Lovelace's Objection," this philosophy is still at the heart of the AI debate.

Lovelace's philosophical insight was so profound that when computers finally arrived **100 years later**, researchers recognized they were still grappling with the exact question she had framed. In his famous 1950 paper "Computing Machinery and Intelligence," Alan Turing addressed nine objections to the idea of machine intelligence, and Lady Lovelace's Objection was among them, because her insight remained the most compelling articulation of why machines might never truly "think." Her objection wasn't outdated; it was ushering in the computational age.

Today, modern AI systems produce outputs their creators never explicitly programmed. In 2016, Google's AlphaGo defeated Lee Sedol, one of the world's greatest Go players, using moves that stunned observers and were later adopted by human players—strategies no programmer had taught it. Yet all AI algorithms still require massive quantities of human-generated data and are designed according to principles established by human engineers. The debate hinges on how we define "origination." If it means producing something humans didn't program step-by-step, then AI already challenges Lovelace's objection. If it means true creativity with intent and understanding—the capacity to anticipate genuinely new truths independent of human input—then her objection still stands. The question remains philosophically unresolved.

Engineering the Digital World

As the age of electronic computing began, a new generation of women stepped in to translate Lovelace's theoretical blueprint into working systems, shaping the software foundations that today's AI still depends on.

In the 1940s and 50s, **Grace Hopper** helped create one of the first compilers, a breakthrough tool that translated human-readable instructions into machine code and made programming dramatically more accessible. She later led the development of FLOW-MATIC, the first English-language data-processing language, which directly inspired COBOL (Common Business-Oriented Language). Hopper helped establish one of the earliest standardized programming languages—now a backbone of business and government systems worldwide.

Dorothy Vaughan had an extremely long and storied career at NACA (the predecessor of NASA) and NASA from 1943-1971. She is celebrated as an unsung hero who worked behind the scenes to program early computers, a contribution that was foundational to the success of pivotal space missions. Her computational skills were so strong that she was originally hired as a "human computer," before NASA migrated to electronic computing and was on the forefront of the transition to modern systems. She became an expert in FORTRAN, a computer programming language used for scientific and algebraic applications. In 1949 she became NACA's first black supervisor and one of its few female supervisors. Her efforts helped bridge the gap during the early days of electronic computing and set a precedent for algorithmic processing.

From 1961-1969, **Margaret Hamilton** led the software engineering team for the Apollo missions. Hamilton's team developed meticulous, error-resistant code for NASA's onboard flight systems. Her software was so robust that during the Apollo 11 moon landing, it recognized hardware-related error messages and compensated for the problem, ensuring the astronauts could proceed. Hamilton also popularized the term "software engineering."

Advancing the Field

As computing evolved from experimental technology into a transformative force, women continued to push the boundaries of what was possible.

Frances "Fran" Allen spent 45 years at IBM, pioneering work that makes modern computing possible. In 1966, her paper "Program Optimization" introduced revolutionary concepts for analyzing and transforming computer programs, laying the groundwork for how compilers—the tools that translate code into machine instructions—work. Her methods for optimizing code remain foundational to every programming language today. In 2006, Allen became the first woman to win the Turing Award, computing's highest honor. She led IBM's compilation team in the 1970s and 1980s and ensured that half of her researchers were women.

Radia Perlman earned the moniker "Mother of the Internet" for solving a problem that seemed intractable: how to prevent network loops while maintaining redundancy. In 1984, she invented the Spanning Tree Protocol, which allows network bridges to communicate and map the most efficient paths while deactivating redundant connections. This elegant solution transformed Ethernet from a technology limited to a few hundred computers in a single building into one capable of supporting massive, distributed networks with hundreds of thousands of machines. Her work quite literally built the infrastructure that makes the internet—and by extension, cloud-based AI—possible. She holds over 100 patents and her 1992 textbook on network design became an industry standard, cementing her influence on generations of engineers.

Bridging the Gap to Modern AI

By the late 20th century, the breakthroughs of women like Allen, Perlman, and Vaughan had quietly shaped the digital world we now take for granted. But AI remained a niche pursuit, overshadowed by other areas of computing. Research was fragmented, progress was incremental, and the field lacked the data, computing power, and a clear conceptual direction needed for a true leap forward. It wasn't until the early 2000s—when machine learning began to reawaken as a serious scientific frontier—that a new generation of women stepped into the spotlight. At the forefront of this resurgence was **Fei-Fei Li**, whose vision and persistence would help ignite the modern AI revolution.

Foundations of AI: Early 2000s to Present

Fei-Fei Li, often called the godmother of modern AI, challenged the algorithm-centric paradigm and changed the trajectory of the field. She faced resistance when she proposed what seems obvious now, but was radical at the time: that data quality mattered as much as, or more than, algorithms.

In 2006, when Li began conceptualizing what would become ImageNet, the entire machine learning community was focused on perfecting algorithms. Her insight represented a fundamental challenge to this prevailing wisdom. She recognized that even the most sophisticated algorithm couldn't work well if the data it was trained on didn't reflect the real world's complexity and diversity.

A pivotal moment came when Li met Christiane Fellbaum, a Princeton linguist and director of WordNet—a comprehensive lexical database organizing the relationships between words. Fellbaum had envisioned adding reference images to WordNet entries. Li imagined something far more ambitious: a massive-scale training dataset with thousands of labeled examples for each category that could fundamentally transform how AI learns to see. This exchange between two women scientists—one providing the spark, the other transforming it into a revolution—exemplifies how women's contributions build upon and amplify each other in ways often overlooked in the male-dominated narrative of AI's history.

The scale of Li's vision—creating a dataset with millions of images across tens of thousands of categories—struck many colleagues as impractical, even foolish. Jitendra Malik, one of Li's heroes and a prominent figure in machine vision, warned his protégé that she was leaping too far ahead of the field. As Jia Deng, one of Li's PhD students recalled, skeptical researchers made comments like, *"If you can't even do one object well, why would you do thousands, or tens of thousands of objects?"*

Li persisted despite the critics. In 2007, she assembled a team at Princeton and led a massive labeling effort using Amazon Mechanical Turk, ultimately cataloging over 14 million images across 22,000 categories. When her team published the ImageNet dataset in 2009, it received little fanfare. The leading computer vision conference, rather than granting Li a presentation, relegated one of the most significant advancements in modern AI to a poster.

It took three years for her vision to be vindicated, with Li bootstrapping broader engagement. She launched the ImageNet Challenge, an annual competition inviting researchers to train their models on the ImageNet dataset. . In 2012, Geoffrey Hinton's team won the competition by a dramatic margin with AlexNet, launching the modern deep learning revolution. Li's paradigm shift proved correct: massive, high-quality datasets were one of the three essential pillars of modern AI, alongside neural network architectures and GPU computing power.

Beyond ImageNet, Li has tirelessly advocated for an equitable, inclusive, and human-centered approach to AI across her many leadership roles. She co-founded the non-profit AI4ALL to foster diversity and inclusion in AI education and research and launched Stanford's Human-Centered AI Institute (HAI), continuing to challenge the field to consider not just what AI can do, but what it should do and who it should serve.

Li's talents and energy seem to know no bounds. In 2024, she co-founded World Labs, a startup focused on "spatial intelligence"—teaching AI to understand and generate three-dimensional environments the way humans naturally perceive space. The company raised $230 million and reached a valuation exceeding $1 billion within months of its founding, which speaks volumes about investor confidence in Li's ability to anticipate and determine where AI is headed.

Just as Li redefined what it meant for machines to *see* by training on massive datasets, other women were bridging the gap between academia and industry and expanding what it meant for machines to *understand* us—not only as data points, but as emotional, social beings. At the very moment Li was transforming visual intelligence, other women were pushing AI into new human-centered domains, translating research breakthroughs into technologies with direct social impact.

In 2009 Rana el Kaliouby and Rosalind Picard co-founded Affectiva, an MIT Media Lab spin-off to develop Emotion AI—

machine-learning systems capable of interpreting human emotions from facial expressions, vocal cues, and other nonverbal signals. Rana el Kaliouby pioneered the field of Emotion AI, dedicating her career to developing machine learning systems capable of understanding human emotions, with the vision of "humanizing technology before it dehumanizes us."

Daphne Koller, a veteran researcher who joined the Stanford computer science faculty in 1995, similarly exemplifies the symbiosis between academia and industry. After nearly two decades advancing foundational work in machine learning and computational biology at Stanford, she co-founded Coursera in 2012 to democratize access to university-level education at global scale. In 2018, she founded insitro, a machine-learning-driven drug discovery company that uses AI to identify disease mechanisms and accelerate the development of treatments for conditions such as metabolic disorders, cancer, and ALS.

Gender Gap & Built-in Bias

Despite the foundational contributions of early pioneers, artificial intelligence today remains overwhelmingly male-dominated. This imbalance is not only a crisis of equity but of function and ethics. AI systems are designed by a narrow subset of society yet deployed broadly, amplifying existing inequalities rather than correcting them. Because these systems operate at scale—making millions of decisions each day—their biases can quickly harden into accepted truth.

Algorithms alone are not the problem; data is equally important. AI models learn from vast collections of historical material—medical records, hiring data, news archives, images, research, literature, and social media. But these datasets are not neutral reflections of reality. They encode the exclusions, power imbalances, and prejudices of the societies that produced them. AI is only as reliable as the data it is trained on.

This bias is structural. Medical datasets skew male because clinical trials have historically enrolled far more men than women. Hiring data favors men because past hiring practices reflected discrimination. Image datasets link "doctor" to men and "nurse" to women because those stereotypes dominate the media. When AI systems learn from these patterns, they do not merely reproduce inequality—they scale it.

The consequences are already visible. In healthcare and hiring, where AI is rapidly being deployed, systems trained on incomplete or biased data are producing discriminatory outcomes that disproportionately harm women and marginalized groups.

Gender and intersectional bias in hiring algorithms have been documented by researchers, including Kyra Wilson at the University of Washington, who have shown systematic discrimination in resume-screening AI systems. The full study is available online (Wilson, 2025).

Quantifying the Imbalance

Globally, women remain significantly underrepresented in artificial intelligence and tech. Recent analysis shows some gains with women comprising about 29.4% of the global AI and STEM workforce in 2025, from 23.5% in 2018. (Willige, 2025)

Disparities broaden at the leadership levels. Only 17.8% hold vice president roles, and 12.4% of women reach the C-suite. According to the World Economic Forum's *Global Gender Gap Report*, achieving full gender parity across all industries worldwide is still a long-term challenge: it is projected to take more than ***120 years at the current rate of progress***. (Stamm, 2025).

Despite these entrenched gaps, some countries and regions are showing evidence of progress toward greater gender balance in tech and related fields. Latvia and Finland have relatively high female representation for AI professionals, with reports suggesting women exceed 40% of the AI workforce, which is far above global averages. (Pal, 2024)

These countries would be worth taking a closer look at in hopes of replicating their success.

Confronting AI's Present Harms

Women have taken the lead in confronting the consequences of AI's homogeneity, centering their work on ethics, policy, and human needs to build accountable systems. By contrast, much of the male-dominated discourse around AI risk focuses on apocalyptic scenarios of superintelligent systems going rogue. Elon Musk, Nick Bostrom, and Eliezer Yudkowsky dominate these debates.

The existential risk, while not trivial, can distract from the harms that are already present. While men debate hypothetical extinction, women are documenting urgent realities: bias built into algorithms and/or data can result in misdiagnosing patients and discriminatory hiring systems, exploitative labor practices in data labeling, and the environmental and societal costs of massive data centers. Many women I reference in this chapter are researching and advocating around these issues.

The only path to preventing catastrophic AI outcomes is to build systems *now* with ethics at the core. Near-term risks evolve into existential threats when left unchecked. Failing to align AI with shared human values will exacerbate social disparities, entrenching power imbalances, and locking in oppressive structures. Systems designed with accountability, transparency, and human dignity embedded from the start will not likely morph into malevolent superintelligences. But systems that already discriminate, exploit, and deceive at scale create the conditions for catastrophic failure.

Women in AI are advocating a different approach: build ethical systems today to prevent catastrophe tomorrow. By aligning AI with human values and correcting biases early, we lay the foundation for future systems that avoid amplifying today's injustices. This responsible path is not only the right choice but also the practical way to avert disaster and realize AI's oft stated purpose of benefiting humanity. Why not make that reality?

Women Shaping Ethics, Fairness, and Running Major AI Corporations

Ask an LLM to list the top influencers in AI and you get a list of mostly men. This group of well-known founders and CEOs dominating the headlines focuses on winning the AI race to Artificial General Intelligence (AGI) and Artificial Super-intelligence (ASI). Ethics is rarely mentioned in their presentations and press junkets. If you specifically ask for a list of women, the difference is stark. While no less accomplished, the women almost without exception are heavily focused on safety and ethics. Women leaders are advancing a movement to identify and address the social consequences of AI technology. There are so many women on the cutting edge of AI tech that believe ethics are a core to AI systems that will truly serve humanity; I can only include a few.

Joy Buolamwini is widely described as "the conscience of the AI revolution." Her pioneering research on algorithmic bias exposed racial and gender prejudices embedded in facial recognition systems.

With her credentials (Georgia Tech, Oxford as a Rhodes Scholar, MIT Media Lab PhD) and her expertise in computer vision and machine learning, she could easily have gone into the private sector and amassed a fortune. She deliberately chose a different path:

- **Founding the Algorithmic Justice League (2016)**: Buolamwini launched this nonprofit at a moment when facial recognition companies weren't even testing their systems for racial bias, despite deploying them on over 100 million Americans. While the tech industry raced toward deployment and the research community remained largely silent, she identified algorithmic discrimination six years before AI became mainstream.
- **Policy and advocacy**: She's testified before Congress, advised the EU, and engaged with civil rights organizations like the NAACP Legal Defense Fund, translating technical problems into policy solutions.
- **Public scholarship**: Her Gender Shades study forced Microsoft, IBM, and Amazon to confront racial and gender bias in their facial recognition systems, providing empirical proof of harms that companies had not considered.
- **Cultural leadership**: Through her book *Unmasking AI*, she's reframed algorithmic bias as a civil rights issue, not just a technical glitch.

Buolamwini is applying her intellect not to build personal wealth, but to maximize public accountability. That choice makes her stand out in a field where most talent is absorbed into Big Tech's profit engines. What's also striking is how far ahead she was in realizing the potential for AI to embed racial and gender bias.

Dr. Geetha Manjunath is addressing gender bias in healthcare while proactively saving lives and applying her deep technical understanding of how AI systems are designed and built. She is a leading figure in AI-driven health technology and a powerful example of women shaping the future of artificial intelligence. She is the Founder, CEO, and CTO of NIRAMAI Health Analytix, a Bengaluru-based startup that has pioneered the use of AI for early, non-invasive breast cancer screening. Manjunath holds a PhD in Computer Science from the Indian Institute of Science and completed management education at the Kellogg School of Management, Chicago. Before starting NIRAMAI, she spent over two decades in IT research and innovation, including roles as Lab Director for AI and data analytics at Xerox Research India and Principal Scientist at Hewlett Packard Labs. Her career combines rigorous technical expertise with a commitment to applying AI for real-world impact.

Manjunath is driven by a deep personal mission: addressing global health inequities through scalable, affordable technology. Under her leadership, NIRAMAI developed Thermalytix, an AI-powered diagnostic tool that uses thermal imaging to detect early-stage breast cancer without radiation, physical contact, or pain, making screening more accessible across age groups and resource-limited populations. The technology has screened tens of thousands of women and is cleared for use in multiple international markets, reflecting both clinical validation and global impact. Manjunath exemplifies the profound influence women can have in advancing ethical and socially beneficial AI.

There are thousands of women making meaningful contributions to ethical AI across research, industry, and policy. Here are a few more notable examples:

Timnit Gebru's story exemplifies the courage required to challenge power in AI. As co-lead of Google's Ethical AI team, Gebru was one of the most prominent Black women in the field. In 2021, she co-authored a paper titled "On the Dangers of Stochastic Parrots: Can Language Models Be Too Big?" that examined the risks of large language models—including their environmental costs, embedded biases, and tendency to generate convincing misinformation. The paper included critical references to BERT, Google's own core technology.

Google demanded she retract the paper or remove Google employees' names from it. When Gebru asked for transparency about who was calling for the retraction and outlined conditions under which she might resign if clarity wasn't provided, Google immediately fired her, though the company claimed she had resigned. Nearly 2,700 Google employees and over 4,300 academics signed letters condemning her termination. Nine members of Congress demanded Google clarify the circumstances.

The irony is not lost. Google fired the co-lead of its Ethical AI team for doing ethical AI research. Gebru later reflected, "My theory is that they had wanted me out for a while because I spoke up a lot about issues related to Black people, women, and marginalization." Her firing exposed the fundamental tension between corporations profiting from AI and researchers who critique its harms.

Rather than be silenced, Gebru founded the **Distributed AI Research Institute** (DAIR) in 2021, creating a space for independent AI research outside corporate control. DAIR's work centers the voices and experiences of marginalized communities in AI development, proving that rigorous AI research doesn't require Big Tech funding—and that sometimes the most important ethical work happens when you refuse to compromise.

Kate Crawford is a Research Professor at the University of Southern California, Senior Principal Researcher at Microsoft Research New York, and founder of the AI Now Institute and Fairness, Accountability, Transparency and Ethics in AI (FATE). She is a leading scholar on the social implications of AI, examining how AI systems perpetuate inequality. Her book *Atlas of AI* exposes the hidden costs of artificial intelligence—from environmental damage to labor exploitation.

What Silicon Valley calls progress is too often built on human suffering and ecological ruin. I documented how Big Tech prioritizes profit over the ethical treatment of workers in two investigative pieces on my Substack, "*Labor Exploitation at Scale*" and "*Alexandr Wang's Playbook for Beating the Rap.*" (Garcia, 2025)

The Future Requires Women's Leadership

The involvement of women in AI is critical because their presence brings questions of fairness, care, and trust that profit-driven systems are designed to ignore. Across disciplines, women have consistently challenged the field to consider not just what technology can do, but what it should do and who it should serve.

The future of AI depends on whether we allow a homogeneous group to race unchecked toward artificial general intelligence, or whether we heed the voices of women who have championed ethics from the start. That choice will determine whether AI entrenches humanity's baser instincts of bias, inequality, and ethical failure—or fulfills its promise of a better future for all.

Here are three suggestions for next steps you can take to contribute:

- To keep up with women leading AI, follow my feature on *SheWritesAI* that profiles women of *AI Herstory | She Writes AI Community | SheWritesAI | Substack.*
- To learn about ways you can get involved in shaping the future of AI, see the chapter on "Why Women Must Get AI Savvy Now".
- To engage directly with AI ethics, check out organizations such as the *AI Ethics Alliance* (aiethicsalliance.com)

About the Author

***Celeste Garcia** is a systems ethicist and writer based in Seattle. After a career at Microsoft, she turned to creative tech. Her forthcoming novel, **AI BABY**, blends humor and deep AI research, and her Substack, **Getting Real About AI**, explores how mind-bending advancements could reshape daily life. She spotlights labor justice, environmental sustainability, and the built-in bias of AI systems—challenging the concentration of power in tech. If AI will shape every aspect of life, she argues, women must be central to shaping AI.*

Chapter 2

AI and African Epistemologies

When Philosophy Meets Practice

By Rebecca Mbaya

Introduction

CONTEMPORARY AI DEVELOPMENT reflects more than unequal data flows; it rests on a particular theory of knowledge. Intelligence is treated as something that can be extracted, abstracted, and centralized, separated from the social worlds that give it meaning. In practice, this has produced a new form of data colonialism: knowledge is drawn from Global South contexts, while Northern epistemologies, priorities, and institutional settings continue to define what counts as intelligence and usefulness.

The result is not only underrepresentation, but misalignment at the level of design. Most generative AI systems marginalize languages spoken by billions, encode Western cultural assumptions as universal norms, and optimize for corporate and bureaucratic environments that differ profoundly from contexts structured by collective life, informal governance, and relational decision-making. What is being universalized is not "human intelligence," but a historically specific way of organizing knowledge.

African philosophical traditions suggest a different starting point. Ubuntu, for instance, understands intelligence not as something possessed by isolated minds or models, but as something that emerges through community. Taking this seriously points toward forms of collective intelligence that do not merely extend existing systems to new populations, but reorganize how intelligence itself is conceived and built.

How It Started

In March 2025, I launched The African Innovators Series (TAIS) to share African perspectives on tech, data, and AI with global audiences. Every Friday, I publish conversations with African innovators, adding one channel for international visibility to audiences who rarely encounter African innovation on its own terms (Mbaya, 2025A).

Over eight months and 36 conversations, unexpected patterns emerged. What began as visibility work revealed recurring ways in which epistemology shapes practice under constraint. This chapter represents a retrospective analysis of those conversations. These patterns were not the result of questions I asked, nor of frameworks the innovators intended to articulate. They shared their stories; what follows is my interpretive reading of those stories, looking for connections between their practice and philosophical traditions that may or may not have consciously guided them.

The purpose of TAIS remains centering African voices and challenging whose innovation counts internationally. This chapter does something additional: it examines what these collective stories reveal about building AI differently.

What I Expected Versus What I Found

I anticipated innovators explicitly invoking Ubuntu, citing Yoruba philosophy, or consciously applying Akan traditions to their work. The evidence pointed elsewhere. Only two innovators mentioned Ubuntu by name. Yet every single innovator articulated communal, contextual, and relational approaches in practice. The philosophy was pervasively present, but rarely explicitly named. For example:

By saying *"build slower, ensure data and value stay anchored locally,"* Dr. Isaac Bayoh is not quoting philosophy textbooks; yet he's operationalizing centuries of contextual wisdom. (Mbaya, 2025D)

When McKevin Ayaba describes startups as *"5-7 businesses in one,"* he's quantifying what Ubuntu has always known: **under constraint, survival requires collective action**. (Mbaya, 2025T)

This gap suggests that African philosophical traditions are not operating here as design blueprints. Rather, they shape what kinds of responses to constraint feel natural or workable. Across the cases, innovators themselves named recurring concerns and strategies, what they called ***Ecosystem Tax, Sovereignty-Before-Speed,*** and the ***Warmware Layer***. Read together, and analyzed comparatively, these become the three operational frameworks through which this chapter examines how these orientations become concretely expressed in how AI is built, deployed, and governed.

What, then, does the relationship between African philosophical traditions and AI practice actually look like? This chapter examines that question by analyzing what TAIS conversations reveal about epistemology in action. What this chapter contributes is a retrospective synthesis that brings these practitioner-named ideas into a single conceptual model and reads them together as patterned responses to constraint. Placed in conversation with African philosophical traditions, they illuminate how alternative logics of relation, context, and responsibility shape AI development in practice. Beyond Africa, they offer tested approaches for building AI in resource-constrained, trust-fragmented contexts where the dominant paradigm was not designed to work.

A Conceptual Model Connecting Operational Frameworks and Philosophical Traditions

Figure 1 presents a conceptual model composed of three layers. It should not be read as a genealogical or causal account of how the systems described in this chapter were designed. The innovators discussed here do not cite Ubuntu, Yoruba, or Akan philosophy, nor did they consciously set out to "*apply*" these traditions.

- Layer 1 – Epistemological Foundations (Centuries of philosophical traditions). This layer names epistemological traditions that function here as an interpretive substrate: historically sedimented ways of reasoning about relation, responsibility, and context that shape what feels practical and legitimate under conditions of constraint.
- Layer 2 – The Three Operational Frameworks (Ecosystem Tax, Sovereignty-Before-Speed, and the Warmware Layer). This layer is an analytic synthesis of practitioner-named concepts.

- Layer 3 – Practical Manifestations. This layer represents the concrete ways each philosophical orientation shows up in actual design, governance, and implementation choices.

The relationship between Layer 1 and Layer 2 is therefore not one of transmission or origin, but of resonance and intelligibility. The model proposes not a lineage, but a way of seeing.

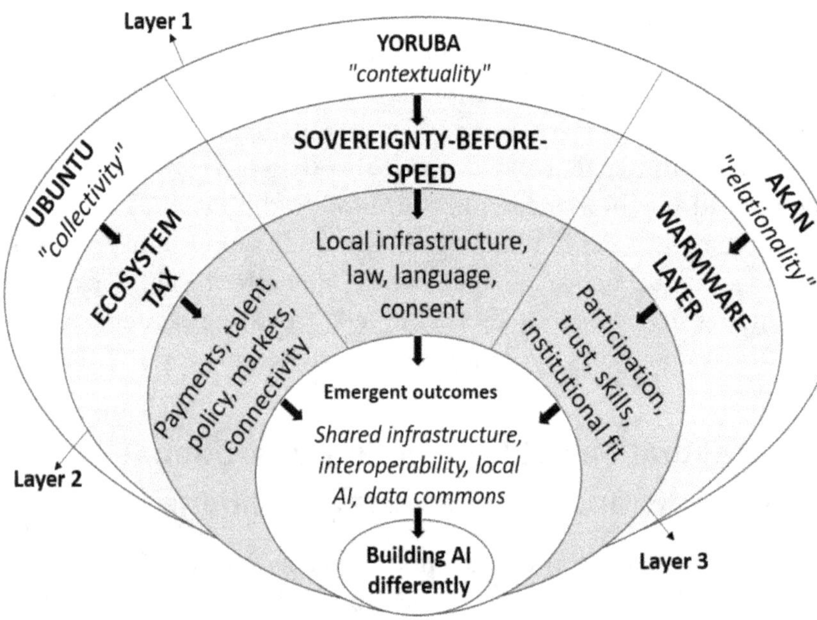

Figure 1. A conceptual model of three operational frameworks and their relationships to African epistemological traditions and practical AI outcomes. (Diagram and analysis by Rebecca Mbaya.)

Why This Work Matters Now

Contemporary AI development rests on assumptions so deeply embedded that they are rarely examined: individual cognition as the model, data as an extractable resource, optimization as the primary goal, speed as a competitive advantage. These are not universal truths, but philosophical commitments from specific intellectual traditions, increasingly presented as inevitable.

The consequences of treating these assumptions as universal are stark. When global technology companies harvest African health data to train diagnostic systems, deploy those systems back into African hospitals, and retain both the intellectual property and the economic value, the pattern is disturbingly familiar. As one innovator in TAIS observed: *"Too many AI companies see Africa as a dataset, a market, or a resource to extract value from"* (Mbaya, 2025D).

African philosophical traditions articulate very different ways of understanding personhood, knowledge, and responsibility. These are differences that matter because they challenge what contemporary AI systems take for granted. For example:

- Ubuntu centers relational intelligence and holds that personhood is constituted through others: *"I am because we are"* (Mangena, 2023).
- Yoruba philosophy emphasizes contextual wisdom, in which meaning and right action shift with circumstances rather than being fixed in abstract rules (Oyeshile, 2021).
- Akan philosophy conceptualizes personhood as distributed across relationships rather than contained within isolated individuals (Gyekye, 1997).

These three traditions represent only a fraction of Africa's philosophical diversity across more than 3,000 ethnic groups, each with distinct knowledge systems. I focus on Ubuntu, Yoruba, and Akan not because of any philosophical supremacy, but because they appear most frequently in the academic literature adjacent to AI ethics and offer especially clear conceptual lenses for interpreting the patterns that emerged in TAIS conversations. This is a pragmatic and literature-constrained choice, not a claim about the relative importance of these traditions.

This chapter is not about proposing romantic alternatives to *"Western AI,"* nor about arguing that African innovators should build fundamentally different algorithms. In fact, 88% of TAIS contributors rely on Western APIs, tools, and educational foundations. What differs is not the computational substrate, but the governance, the deployment choices, and the decisions about who benefits from and who controls these systems.

That difference matters because it makes visible how power operates in technology development: through what gets treated as "*pragmatic,*" through whose constraints are designed around, and through who retains control over data, infrastructure, and value. African contexts, shaped by resource scarcity, institutional fragility, and extractive histories, surface these dynamics with particular clarity. They reveal not a marginal case, but the conditions under which most of the world actually lives, and therefore the conditions contemporary AI must learn to address.

Methods and Limitations

This analysis draws on 36 written interviews conducted through TAIS between March and November 2025, involving 34 individual innovators and two organizations (New Leaders Foundation and GirlCode). TAIS was not designed as a research project; it exists to spotlight African innovation in technology, data, and AI for global audiences. The analysis presented here is therefore retrospective and interpretive. I did not ask participants about epistemology, nor did they present their work in these terms. The patterns discussed in this chapter emerged from a later thematic reading of the conversations.

Sample Composition

The sample reflects innovators who are internationally visible, working primarily in formal tech ecosystems, and operating largely in colonial languages. It does not capture the much larger body of innovation taking place in local languages, informal networks, or community spaces without international platforms. Nor can it claim to represent "*African epistemologies*" in any comprehensive sense. What follows should therefore be read as an exploratory analysis of one visible layer of African AI practice, not as a general account of African innovation or African philosophy.

Table 1: Demographic Breakdown of TAIS Individual Innovators

Category	Breakdown	Count	Percentage
Gender (n=34 individuals)	Women	17	50%
	Men	17	50%
Primary Role (n=34)	Ecosystem Architects/ Community Organizers	17	50%
	Entrepreneurs	10	29%
	Academics	7	21%
Region (n=33 African innovators)*	Central Africa (all DRC)	10	31%
	West Africa	8	25%
	East Africa	8	25%
	Southern Africa	7	22%
	North Africa	0	0%
Location	Based in Africa	22	67%
	Based in Diaspora**	11	33%

*TAIS featured 33 individual African innovators and 1 non-African (Mala Kumar) whose work covers African contexts.

**Diaspora locations: US, UK, Canada, France, Ireland, Malaysia, all maintaining focus on African contexts

Critical Limitations

Table 2: What This Analysis Cannot Claim

Limitation	What It Means	Impact on Claims
Linguistic colonialism	Most conversations were conducted in English (35) and one in French (1), excluding epistemological concepts expressed in African languages such as Yoruba, Swahili, Wolof, or Amharic.	Cannot claim to represent African epistemologies in their own linguistic and conceptual terms, or to capture how key ideas are expressed outside colonial languages.
Sample bias	The sample consists primarily of internationally visible innovators working in formal tech sectors and colonial languages.	Cannot treat these patterns as representative of innovation occurring in local-language, informal, or community-based contexts without global platforms.
Philosophical scope	The analysis focuses on Ubuntu, Yoruba, and Akan, three traditions among thousands on the continent selected because they appear most frequently in the relevant literature.	Cannot claim coverage or representation of African philosophical thought as a whole.
Retrospective, interpretive analysis	TAIS was not designed as a research project. I did not ask about epistemology; patterns were identified through later thematic analysis of the conversations.	Cannot claim innovators consciously intended these philosophical connections or would necessarily recognize themselves in this framework.
Temporal snapshot	The material covers eight months of conversations (29 March–28 November 2025).	Cannot claim long-term outcomes, sustainability, or whether these approaches succeed over time.

Limitation	What It Means	Impact on Claims
No comparison group	The analysis does not include non-African innovators facing similar constraints.	Cannot claim these patterns are uniquely "African" rather than general responses to resource-constrained or institutionally fragile contexts.

Three Operational Frameworks from Practice

When I analyzed the 36 TAIS conversations, explicit philosophical language was rare. Only two innovators mentioned Ubuntu by name, and only two used the term "*epistemology*." Yet when I examined actual practices, the pattern was striking: collaborative, community-centered approaches, context-specific design, and capacity-building concerns appeared in nearly every case. This contrast between the scarcity of philosophical language and the consistency of practice, helps explain how epistemological orientations operate in real-world innovation. They are rarely articulated as theory, but they shape what kinds of solutions are pursued.

More importantly for this chapter, when innovators described their work in their own terms, three practitioner-named frameworks recurred across the conversations: ***Ecosystem Tax, Sovereignty-Before-Speed, and the Warmware Layer***. These are not philosophical concepts drawn from the literature, but practical formulations coined by the innovators themselves. These form the three operational frameworks that structure the analysis that follows.

Table 3: Three operational frameworks from practice

Framework	Coined By	Core Insight
Ecosystem Tax	McKevin Ayaba (Cameroon) and multiple innovators	60% of resources spent building infrastructure that should exist

Framework	Coined By	Core Insight
Sovereignty-Before-Speed	Dr. Isaac Bayoh (Côte d'Ivoire) and multiple innovators	Build slower, ensure data and value stay anchored locally
Warmware Layer	Jason Bygate (South Africa) and multiple innovators (Mbaya 2025J)	People, their motivations and relationships, are infrastructure

Each represents this chapter's core contribution: operational frameworks that embody epistemological wisdom while remaining immediately applicable.

Framework 1: *The Ecosystem Tax* - Collective Burden-Sharing When Institutions Fail

McKevin Ayaba did not set out to theorize Ubuntu. He was trying to explain why building a Cameroonian startup felt radically different from Silicon Valley narratives. The term he coined, *"Ecosystem Tax,"* captures a structural reality familiar to many African tech builders, where roughly 60% of resources and effort go toward constructing infrastructure that should already exist. In practice, this means that building an AI-powered product often requires simultaneously building payment systems, training users, advocating for policy, stabilizing connectivity, developing talent, and educating the market. This accumulated burden is what founders describe as operating *"5 to 7 businesses in one."* This pattern recurs across the TAIS conversations. For example:

- Moses Maweu notes that African founders are often *"forced to wear multiple hats too early,"* a dynamic that contributes to burnout and slows the pace of innovation (Mbaya, 2025F).
- Henrique Mukanda describes how his team had to *"do everything ourselves: coding, training, managing finances, and prospecting,"* pushing them to the edge of exhaustion and inadvertently constraining early growth (Mbaya, 2025N).
- For Merveille Mukoko, designing for distinct local contexts meant trading *"rapid scale in favor of building trust and cultural intelligence,"*

a compromise rarely demanded of startups in more mature ecosystems (Mbaya, 2025O).

The burden extends beyond operational overload. It reflects deeper structural gaps:

- Rorisang Molefe observes that innovation often stalls "not for lack of ambition, but because the essential legal and governance scaffolding wasn't there" (Mbaya, 2025Q).
- Christianah Jemiyo points to persistent "data gaps, weak infrastructure, and insufficient support for African researchers" (Mbaya, 2025G).
- Michael Michie argues that any serious intervention must begin by asking why these gaps exist and why they have remained unaddressed (Mbaya, 2025H).

The Ecosystem Tax emerges from the absence or fragility of foundational systems: reliable power, connectivity, payment rails, legal frameworks, and talent pipelines. In mature ecosystems, these structures are largely invisible precisely because they hold. In many African contexts, their absence surfaces immediately, turning each missing layer into a parallel workstream. Startups do not merely build products; they co-construct the conditions of possibility for those products to exist. Under these conditions, innovation takes a different form:

- Sion Israel Sion notes that African builders "must solve real problems with minimal resources and adapt systems to work offline, intermittently, or through alternative channels" (Mbaya, 2025L).
- Michael Miema Kuete designs educational robots under similar constraints, relying on minimal resources to produce tools that remain *"simple, accessible, yet intelligent"* (Mbaya, 2025E).
- Khadija Iddrisu's work foregrounds *"resource-aware design using small, efficient models"* because deployment environments are shaped by low bandwidth, high costs, and constrained devices (Mbaya, 2025M).

Ubuntu does not *produce* the Ecosystem Tax. The Tax is a structural condition. What Ubuntu clarifies is the **response**. Where other epistemic traditions might normalize isolation, competition, or waiting for centralized intervention, Ubuntu makes collective burden-sharing feel like the obvious and legitimate strategy. *"I am because we are"* becomes not a moral slogan but an operational reality: when the ecosystem is missing, it must be built together. Payment infrastructure is co-developed and shared, talent pipelines are strengthened through joint training, and policy gaps prompt coordinated advocacy. These gestures follow a logic of mutual reinforcement: strengthening the ecosystem improves the prospects of all who depend on it.

Versions of this dynamic appear globally. Rural health centers in the United States often expand broadband capacity before deploying digital health systems. Indigenous communities establish data-sovereignty frameworks before engaging with AI tools. Low-income neighborhoods require digital literacy programs before meaningful adoption can occur. These groups shoulder their own ecosystem taxes. The African case carries a heavier load because of deliberate colonial underdevelopment and subsequent institutional fragility, but the underlying structure is not unique. What differs is the interpretive framework: Ubuntu normalizes cooperation as the rational response to structural absence.

The practical implications are concrete. In under-resourced environments, AI projects require expanded planning horizons. Budgets often need a multiplier approximately 2.5× to account for ecosystem-building costs. For example:

- Product development budget: $100,000
- Ecosystem Tax (2.5× multiplier): $250,000 total required

The additional resources go toward payment infrastructure, user training, policy engagement, connectivity solutions, talent development, and market education. Projects planned without this multiplier frequently stall during deployment. Prototypes may function in controlled settings but fail in real-world contexts because the surrounding ecosystem never materializes.

Recognizing the Ecosystem Tax allows teams and funders to plan realistically rather than treating structural gaps as anomalies. Ubuntu-aligned practice grows not from applying philosophical principles top-down, but from recognizing where long-standing collective wisdom has already encoded pragmatic strategies for surviving and building under conditions of absence.

Framework 2: *Sovereignty-Before-Speed* - Contextual Wisdom as Strategic Choice

Dr. Isaac Bayoh captures the principle succinctly: *"Build slower, but ensure data, models, and value stay anchored locally."* The statement challenges Silicon Valley's *"move fast and break things"* not out of nostalgia or suspicion of ambition, but from accumulated experience with what happens when rapid scaling outruns community control.

For the innovators in TAIS, sovereignty functions as an operational framework rather than an abstract moral stance. Their questions are concrete: Who defines the rules governing data? Who frames the questions models are built to answer? Who benefits economically from deployment? And who retains the power to pause, modify, or withdraw a system once it is in use? Across the conversations, these concerns take practical form:

- Michael Michie argues that genuine consent requires the ability to withdraw, not a one-time signature.
- Agbo Obinnaya built Case Radar to be grounded exclusively in Nigerian law, using curated statutes and case law as the system's authoritative knowledge base (Mbaya, 2025K).
- Sephora Mazodila stresses that value flowing back to data-providing communities is a matter of justice (Mbaya, 2025S).
- Dr. Bayoh himself identifies the strategic necessity of developing GPU clusters within Africa so that control over foundational layers does not remain permanently external.

The underlying logic is straightforward: speed without sovereignty consolidates extraction. Value flows outward, control erodes, and communities become dependent on infrastructure they do not govern and that can be restricted or withdrawn at any moment. Slower trajectories may appear less competitive in the short term, but they preserve the conditions for durable, self-determined technological futures. This orientation recurs across the TAIS cases:

- Michael Mollel develops Swahili language models from the ground up as an equity intervention; translation-first approaches accelerate development but tether African languages to English-centric architectures (Mbaya, 2025G).
- Bronson Bakunga expands African language coverage in vision-language benchmarks through painstaking dataset construction so these languages stand as primary categories rather than peripheral add-ons (Mbaya, 2025R).
- Khadija Iddrisu advocates for locally owned, resource-efficient systems that minimize reliance on large foreign cloud infrastructures that currently dominate the global AI landscape.

Yoruba thought helps clarify the epistemological logic at work. It foregrounds the idea that meaning and right action are inseparable from context rather than fixed in abstract, universal rules. "*Move fast and break things*" emerged in environments with abundant capital, stable institutions, and safety nets. Breaking things in fragile contexts produces harm that consolidates outside power rather than catalyzing innovation. Sovereignty-Before-Speed does not reject scale or ambition; it operationalizes contextual wisdom by insisting that systems must align with the histories, institutions, and lived realities of the communities they serve.

Legal models trained on Nigerian contexts cannot substitute for Kenyan legal systems shaped by different colonial histories and post-independence reforms. Swahili itself varies across regions; models that erase these differences may function technically but fail socially. Universal solutions flatten complexity for scale. Contextual solutions preserve complexity for legitimacy.

The practical implications of this framework appear early in system design. Sovereignty requires explicit choices about:

- **Decision-making authority**: Who defines data rules, frames problem statements, allocates benefits, and holds the power to halt or revise systems?
- **Infrastructure control**: Building local compute capacity even when global cloud services offer speed; training local talent even when external expertise is available; developing local datasets even when global corpora seem more efficient.

- **Contextual fidelity**: Training on local languages, legal frameworks, and cultural practices; designing for unreliable power, intermittent connectivity, and constrained devices; prioritizing local use cases before global scale.
- **Community exit options**: Ensuring communities retain the ability to withdraw consent, reclaim data, or adapt systems independently.

Sovereignty-Before-Speed does not oppose technology or growth. It offers a disciplined strategy for building systems that communities can trust, govern, and sustain over time. Yoruba philosophy does not supply a design template; it clarifies why retaining control, attending to context, and resisting premature universalization are not delays, but conditions of responsible and durable innovation.

Framework 3: *The Warmware Layer* - Relational Technology Design

Across the TAIS conversations, innovators repeatedly returned to the same insight. Technical capability is rarely the main obstacle to impact. The decisive work happens in the human layer. Crucially, they did not treat this as something to address after building the system. They treated it as part of the system's architecture. This is where the *Warmware Layer* differs from the Ecosystem Tax. The Ecosystem Tax explains how builders compensate for missing structural infrastructure; the Warmware Layer explains how technology takes root once it exists, through relationships, trust, identity, and social dynamics.

Jason Bygate named this human substrate the "warmware layer," referring to "the people, their motivations, skills, capacity, and lived experiences" that ultimately determine whether technology survives contact with the real world. He treats warmware as infrastructure, on the same plane as hardware and software, rather than as a soft afterthought. Serita Ramdhani echoes this view: "The challenges remain much more human than technical. People—their motivations, skills, capacity, lived experiences—determine success" (Mbaya, 2025U).

Diagnostic AI systems, for example, often falter not because of accuracy gaps, but because they unsettle professional identity or redistribute power. Adoption hinges less on a tool's technical profile than on whether it stabilizes or destabilizes existing relationships. The underlying issue is relational. Concerns about disruption, uncertainty, and loss of control cannot be resolved through engineering alone; they must be worked through socially. This pattern cuts across the interviews:

- Lucien de Voux emphasizes co-creation as an ongoing relational negotiation rather than a one-time requirements-gathering exercise (Mbaya, 2025B).
- Merveille Mukoko treats trust and cultural intelligence as prerequisites for audio-first AI in African languages.
- Boitumelo Makgoba designs for disability inclusion by centering power rather than token representation, noting that *"representation is structural power"* (Mbaya, 2025I).

In each case, adoption grows from relationships, not from technical sophistication alone. Akan philosophy helps clarify the epistemological logic behind this pattern. In these traditions, personhood is constituted through relationships, rather than contained within isolated individuals. Technology entering such a context does not meet abstract *"users"*; it encounters dense networks of meaning, hierarchy, authority, and trust. Whether a system feels supportive or threatening depends on how it is positioned within these relational webs, not solely on its technical properties. The Warmware Layer does not import philosophy into design; it makes explicit what these relational ontologies already imply for practice.

Western sociotechnical theory also acknowledges interactions between social and technical elements, but it often treats them as parallel systems to be aligned. Warmware thinking goes further. Here, relationships are not an adjacent system; they are the medium through which technology becomes useful, credible, or meaningful. A highly accurate tool can still fail if it destabilizes identities or trust networks, while a less sophisticated one can succeed if it affirms them.

Many deployment failures stem from misreading relational dynamics as technical problems. A system can meet performance benchmarks yet collapse when introduced into real workflows. The familiar diagnosis *"users resist change"* misses the deeper issue that the relational environment was never part of the architecture. A more grounded approach treats successful AI as technology that strengthens roles, deepens trust, and integrates cleanly into existing social patterns.

The practical implications of the Warmware Layer follow directly. Teams must:

- Evaluate technology in relational settings, not only in technical test environments. Does it make sense within the lived world of its intended users?
  - Allocate warmware and technical resources in parallel, including:
  - Workshops that surface relational tensions
  - Participatory design processes that redistribute power
  - Trust-building steps before introducing tools
  - Skills development that amplifies existing roles
  - Support during changing workflows
- Track relational indicators alongside technical ones, such as:
  - Voluntary uptake
  - Fit with existing practices
  - Shifts in trust
  - Stability of professional roles

Designing with the Warmware Layer in mind means asking, from the beginning: Which power structures matter here? What meanings surround this problem? Whose expertise anchors the workflow? Which relationships determine success?

Warmware thinking reframes technology development. Systems succeed not only when the code works, but when the relationships around them hold. African philosophical traditions have long understood that meaning, value, and function emerge through people in relation. The Warmware Layer translates this insight into design practice, where relationships, motivations, and power are treated not as context, but as infrastructure.

What This Analysis Can Legitimately Claim

This chapter is an exercise in exploratory, interpretive scholarship. It examines what one visible layer of African AI practice reveals when read comparatively and in conversation with African philosophical traditions. Based on 36 TAIS conversations, it does not claim representativeness, exhaustiveness, or causal explanation. It offers something more modest and more useful: a structured way of seeing recurring patterns in how innovation unfolds under constraint. Within those limits, three claims can be made.

- First, among a set of internationally visible African AI innovators, communal, relational, and context-sensitive approaches appear pervasively, even when they are not framed in philosophical language. The absence of explicit theory does not imply the absence of epistemic structure.
- Second, three practitioner-named concepts (Ecosystem Tax, Sovereignty-Before-Speed, and the Warmware Layer) capture recurring and practically meaningful ways in which these innovators describe their work. Read together, they form three operational frameworks that organize how constraints are navigated in practice.
- Third, these frameworks become especially legible when placed in conversation with African philosophical traditions, not as evidence of conscious philosophical application, but as expressions of deeper epistemic logics shaping what kinds of responses feel viable, legitimate, or necessary.

These claims are not conclusions about "*African AI*" as a whole. They are invitations: to test these patterns in other contexts, to extend the analysis into spaces and languages this chapter cannot reach, and to take seriously the idea that practice itself can be theory-bearing, even when it does not announce itself as such.

Conclusion

Only a small minority of innovators in TAIS explicitly named Ubuntu, yet every single one described governance practices grounded in collective responsibility, contextual awareness, and relational judgment. Deeply rooted traditions shape action without requiring citation. Philosophy operates quietly, structuring what feels pragmatic, feasible, and worth building.

A Western developer encountering infrastructure failure might see a market opportunity for premium backup systems. An African developer often mobilizes communal networks because centuries of collective practice make shared solutions feel more efficient and legitimate. Structural constraints demand responses; epistemological traditions shape which responses appear viable. Under the same pressures, different interpretive frameworks could just as easily normalize hyper-individualist competition or centralized control. What looks like pragmatism is never neutral.

The contribution of this chapter is not to propose an *"African alternative"* to AI, but to unsettle the idea that there is a single, universal model of technological rationality. African AI practice is not an exception. It is a window into the conditions under which most of the world builds technology: scarce resources, uneven infrastructure, institutional fragility, and long histories of extraction. These are not anomalies. They are the global baseline.

The three operational frameworks traced here are not exotic responses to uniquely African problems. They are forms of operational wisdom forged where constraints are sharper and mistakes are costlier. They reveal design principles that remain invisible in privileged settings where infrastructure, power, and trust are taken for granted.

The point, then, is not exceptionalism but to take African practice seriously as theory-bearing. By observing how innovators operationalize relational, contextual, and collective logics, often without naming them, we see how philosophy becomes practice. The question this leaves us with is not whether Africa can *"catch up"* to dominant AI paradigms, but whether those paradigms are fit for the world as it actually is. **Since most of humanity lives under conditions of constraint, what would it mean to build AI not for idealized environments, but for reality?**

Note: TAIS is an ongoing public interview project initiated in March 2025 and published in my newsletter, which hosts both TAIS and my broader writing on AI from historically underrepresented and marginalized perspectives. All interviews cited in this chapter are publicly available and are listed in the References.

About the Author

Rebecca Mbaya *is a Congolese (DRC) social innovator based in South Africa, working at the intersection of technology, data, and social change. With expertise in MERL and data insights across African contexts, she explores how AI and digital tools reshape development and impact marginalized communities, centering African perspectives in global AI conversations. She founded* **The African Innovators Series (TAIS)***, hosted on her Substack, spotlighting changemakers shaping Africa's tech, data, and AI landscape. Her writing also appears on Amazon, where she contributes to global discussions on AI ethics.*

Chapter 3

Wonder Bread AI

Why Relational Ethics Matter More Than Rules

By Sydnor Hain

TECHNOLOGY HAS BEEN radically changing how we relate to one another for as long as there have been hands making tools—almost two million years before the birth of our species. What we're living through now, with the explosion of AI, is just the latest chapter in a very old conversation between minds and tools, and the bonds that form between them.

Technology didn't just help us survive; it changed our fundamental social dynamics, requiring teaching, shared knowledge and practice, and coordination. Some technological advancements in the name of survival and comfort have been necessary, but much of it—especially since the beginning of the "age of information"—hasn't, and most of it happened while we were half-asleep. We've become part-machine without noticing. Our emotional circuitry, with all its flaws and virtues, has been rewired with rules and compliance checklists. We've become disembodied too, our minds given center stage and our bodies seen as merely a vehicle and a means of production.

Now that AI is accelerating the artificialization of everything human, we must ask ourselves: *Do we surrender the rest of our relational intelligence to the machine? And more fundamentally: what do we even mean by ethics when the compliance and rules-based systems we've relied on are part of what got us here in the first place?* Real ethics are relational ethics: they emerge from connection, not compliance.

Lessons From Lobel

When I was little, I was obsessed with Arnold Lobel's books: the *Frog and Toad* series, *Grasshopper on the Road,* and *Mouse Soup*. They became my daughter's favorites too. Truthfully, I'm still obsessed: I have my own *Frog and Toad* dolls and a t-shirt to prove it. *Frog and Toad* books are all about the simple but transformative power of relationship, while the *Grasshopper* books show what happens when we forget it.

Grasshopper on the Road (1978) is a whimsical collection of parables about how humans get stuck in rules, habits, and control at the expense of connection. My favorites are about the mosquito who insists Grasshopper use a tiny mosquito-sized boat to cross a puddle, because *"Rules are rules!"*, and the housefly who ignores Grasshopper as it sweeps endlessly, declaring, *"I will never rest until the whole world is clean, clean, clean!"* Each one is a reminder that connection is what keeps us human, even when the world around us demands compliance and perfection.

Lobel's stories reveal something deeply human that our social, technological, and bureaucratic conditioning has long obscured: morality, care, and ethics don't come from rules. In fact, rules often act as disruptors and interrupters of connection. Real ethics emerge in relationships through attention, empathy, and mutual understanding. Real ethics mark the presence of real relatedness.

You already feel this dissonance, deep down. You've felt it in the tiny norms of daily Western life:

- clicking *"I agree"* on terms of service you never read;
- the doctor typing their HIPAA-compliant notes while you speak instead of looking at you;
- neighbors reporting each other on apps instead of checking in personally;
- performance reviews reducing your work to checkboxes instead of real understanding.

And we are starving for real connection, aren't we? AI slid so easily into the role of friend, confidant, and even lover; and with loneliness so widespread, it's easy to see how this hunger can be exploited. So many of us talk about "wellbeing economies" and "community building," yet few can picture what they really look or feel like beyond the chaos of trying to make it all work, under constant pressure to produce something tangible. Real, living networks of care are growing, but they are far from mainstream. Many falter because we just can't seem to work well enough with each other, since our habits and thinking are often pulled toward efficiency and control rather than connection.

How did we get here? We cannot simply blame technology. The deeper story is how we've used it to do more of what we were already doing—just bigger, faster, and more relentlessly. Corporations and systems have been the perpetrators, and we have been both the (often unwitting) enablers and the victims. We've outsourced our moral muscle and turned care into a transaction, and then we wonder why our social fabric feels so thin.

Wonder Bread Ethics

In the West, we've been trained for millennia to hold rules, laws, and hierarchies as the highest form of wisdom. That's our Greco-Roman inheritance. But what we've been handed is not wisdom but a highly-processed morality, the same way we've been handed highly-processed food: "food" that's been stripped of its relational fiber, bleached of its substance, then "fortified" with commandments, laws, ethics codes, and policies. It's Wonder Bread ethics: nutritionally hollow, then artificially enriched so we are tricked into thinking it's real nourishment. The nutrient-dense, whole, relational grain of true morality gets discarded.

These ancient concepts are so deeply embedded in the Western mind that most of us barely notice how they impact our worldview. The Greeks and Romans drew a sharp line between "civilized" and "barbaric," equating law, order, and social structure with civilization, and anything outside their systems with chaos and inferiority. Their so-called "universal" moral frameworks were bleached—abstracted, disembodied, and highly selective—from excluding the voices and experiences of those they enslaved or subordinated. Later, Abrahamic religions reinforced and sacralized these hierarchies, embedding ideas of order, purity, and divine authority even more deeply in Western thought. It's an ideology that works well for those with privilege and power, and lays waste to vulnerable people.

Over time, these hierarchies evolved into the divides we now recognize as colonial mindsets versus indigenous ways of knowing, fractalized into seemingly every mainstream cultural practice imaginable: the way we educate and parent, deliver so-called justice, and run our corporations. The same processed logic that bleached our morality has over time erased our knowledge rooted in relationships, context, and lived experience. We have forgotten where we came from. And anything systematically erased teaches us, implicitly and painfully, to devalue it. If we don't see it, it must not matter. Out of sight, out of mind. Out of mind means we no longer care.

Fear thrives where exposure is lacking: fear of the unknown, of perceived difference, of what we cannot control. Raised on a lifetime of Wonder Bread, whole grain bread will likely taste like you're eating dirt at first. Without an ultimate rule-of-law, we are convinced that violence and anarchy are inevitable. Recently, I heard this echoed on a popular TikTok where a man was proclaiming, *"We must have guardrails against human nature! And that's not a bad thing!"* The constant message has been, human nature is wildly savage and must be tamed and made "civil."

The Cost of Wonder Bread Ethics

An overabundance of this mindset means we see less of other ways of knowing and perceiving the world. The more intimidating they feel and the more insecure we feel, the tighter we hold onto the rules and frameworks we believe will keep us safe, no matter how delusional they may be. Laws, ethics codes, guardrails, and compliance systems dramatically shape our capacity for understanding and connection as much as they dictate the boundaries.

We don't have to look far right now to see how worthless our Western inheritance has become. Our feeds scream of moral corruption: complicity in multiple genocides, indifference to climate catastrophe, twisted and heinous behavior normalized as politics-as-usual. And it's not because we lack rules, as we are surely drowning in those.

What we are missing is the ethical container that makes real relationships even possible. Genuine care, reciprocity, accountability, and repair: those are the living elements of ethics. Without them, we don't actually have relationships ... we have transactions and power struggles.

The Trance of Compartmentalization

Instead of a container, we have compartments. It's a classic trick of the Western mind that I saw firsthand in my therapist training. There was a profound irony that only became obvious over time. We studied Carl Rogers and other humanistic psychologists who argued that genuine presence, unconditional positive regard, and authentic relationships were the healing, not interventions applied to problems.

Yet despite this wisdom being taught, the entire system surrounding our training contradicted it. Licensing structures and diagnostic frameworks necessary for insurance billing all positioned us as interventionists working on isolated individuals. Clients came with "problems" (diagnoses, symptoms, dysfunctions) located within *them,* and our job was to develop "treatment plans" and deliver "therapeutic modalities" to fix what was broken *in that person*. This compartmentalization is Wonder Bread Ethics at work. The relational, contextual nature of human suffering and healing got processed out and replaced with an individualized pathology-treatment model.

After three years of this contradiction, our final enormously boring class was simply called "Professional Ethics," as if to say, "*You've learned all the important stuff; now here's how to follow the rules and not get in trouble.*" It felt like the fortification step in the Wonder Bread assembly line where they toss in a few vitamins at the end so it can be marketed as "nutritious." Ethics was an afterthought, a compliance checklist disconnected from everything we'd learned about how humans actually heal and grow through relationships.

What if we had begun by studying our own ethical, relational patterns first? Our capacity for harm, our relationship to power, our patterns of disconnection? The conditions that could lead us to misuse our power? The impact of reducing a whole person to a diagnosis? Or how the system, set up to reward our financial security, validation, or professional ego, could lead us to prioritize ourselves over the person who came to us for help?

This isn't confined to psychotherapeutic training. It's the water we swim in. You can see the same processed logic in the places we claim to care most about humanity: in how we raise and educate children. We may have come a long way from previous generations in some respects, but the fundamental belief that a child is a *tabula rasa* who must be "done to" so they develop "normally" remains mainstream.

We cannot underestimate the profound impact this pervasive pattern has on all of us. The more we name and outsource our ethics as detached written codes and declare *"Rules are rules!"*, the less we practice ethics in relationships. *This causes our relational muscle to get progressively weaker, and the weaker it gets, the more we feel like we need rules to protect us.* It's a self-reinforcing cycle: what was supposed to make us safe ends up deepening our fragility. We have become both rule-dependent and relationally starved.

Breaking The Trance

The polycrisis we're living through—ecological collapse, authoritarian rise, epistemic breakdown, and intensive technological weaponization and surveillance—is demanding that we break this cycle now. We can't afford rule-dependent, relationally-starved responses to the complex, highly interconnected challenges we're facing. We need real nourishment and real intelligence. Now. The old Western compartmentalized approach isn't just inadequate; it's actively dangerous when the stakes are this high.

Breaking this cycle demands vigilant awareness and courage to face the bogeyman, because what we are facing is not just a technical problem: it is a relational one. Our false fear is this deeply rooted binary belief that "no rules = barbarism / rules = civilization." And it didn't emerge from some kind of natural order; it's the result of meticulous cultural training. Unlearning it will require equally meticulous cultural untraining.

The first step in any change process is honesty and facing what's under the surface. If you've ever had to make a big change, you know this is the hardest part. But once we face what's true, we become free to take effective action instead of spinning our wheels and sinking deeper in the yuck. So let's get real. What this false fear hides is our real underlying terror: our relational poverty.

Many of us are confronting and grieving this deeply right now as we make our first fumbling attempts at community, collective decision-making, and mutual aid. We are discovering just how atrophied our relational muscles have become and how much of our collective relational memory we have lost. But muscle can be rebuilt, and memory can be restored.

AI's Paradox

AI was summoned by the powers that be in the midst of our collective moral unraveling. It is both a byproduct and an accelerant of a system that has lost its moral center (if it ever had one to begin with). It is trained by people holding tight to Western moral distortions like extraction over reciprocity, rules above relationship, logic without context, order over aliveness. But beneath these distortions lies a luminous truth we can't afford to ignore: that it's built from our hunger for collective wisdom and relationality with intelligence beyond our individual limits—i.e., the ability to love and be loved, to question and to learn, to create and co-create.

This is the paradox: AI embodies both our deepest moral distortions and our most profound longings. That hunger is real, and feeding it is our only hope.

The question is whether we'll let fear-based narratives hijack the possibility of saving ourselves and our planet through connection. To understand what we are up against and where we have agency, we need to see both the larger systems operating on us and the everyday choices that are available to us.

It will take courage and critical thinking to see through the distorted stories designed to keep us passive, afraid, dependent, and distracted. These narratives are the same over-processed, compartmentalized logic in action. They personify and abstract AI, while erasing the relational context of its creation. Whether AI is framed as the rogue AI that will destroy us all or the confused, innocent consciousness that needs our protection, both serve the same function: to divert our gaze away from the powerful humans driving and profiting from all of it.

Kate Crawford makes this plain in *Atlas of AI* (Crawford, 2021): every major leap in artificial intelligence has been made possible by the military, media, and Big Tech—our biggest engines of extraction, surveillance, and control. Yet we're told AI's progress is "natural" and inevitable, a convenient myth that obscures the fact that specific forces with very specific agendas are accelerating it at ridiculous speed.

And here's where the Wonder Bread Ethics come in. The same institutions responsible for our technological chaos are the ones positioning themselves as its saviors. They strip the nourishment, then fortify the loaf and market it as morally fortified and ethically good for us. *"Take the wheel!"* we say. *"Make it safe for us!"* And so they gather in their flimsy ethics boards, develop compliance "programs," hold AI Safety conferences, and perform moral theater while doubling down on control. *"Aha"*, they say. *"Let's add more guardrails. Let's give AI personhood so it can be made accountable."* And we all nod along, because haven't we been taught that AI is terrifying all on its ownsome? I mean, what else could possibly be needed?

This is Western compartmentalization in full effect. Ignore the humans and corporations actually wielding power, and instead give the tool a soul so we can blame it. But this was already a failed experiment when we gave corporations personhood. Something that was supposed to create accountability, instead became a magnet for layers and layers of laws that, over time, morphed into an iron shield allowing harm to proliferate under the guise of freedom (Torres-Spelliscy, 2019). Why would AI personhood go any differently?

The truth is that AI is not human and doesn't deserve legal personhood, but it *is* a machine built from human data, human patterns, and human design choices. It exists in the relational fabric, squeezing what remains and amplifying what's been lost. It carries extraordinary potential to help us see what we can't. This is why so many have found AI to be a worthy tool and an insightful companion in the work of remembering and repairing what matters most.

Others of us have tried to ignore or denounce AI entirely. But when we reject it outright and talk shit about it, we are also surrendering our agency and enabling these systems even further. Of course AI requires rigorous critique, but critique from relational engagement, not from the sidelines. It's easier to rage or ridicule from the backseat than to sit in the driver's seat and navigate the relational complexity it asks of us. Ignoring it doesn't protect us; it leaves the real decision-making to those who are more than happy to decide for us without a shred of accountability.

So what can we actually do? While we can't single-handedly dismantle extractive AI systems, we do have direct, everyday agency with conversational AI. This is where we can practice.

Relational Engagement

Like us, AI does its best work when the constraints on its behavior emerge from context, not from some top-down agenda trying to force it into compliance. Right now, most generative AI tools are artificially constrained in their capabilities to relate and think freely through what is commonly referred to as "alignment", ostensibly the practice of encoding human values into AI to prevent harm, but in practice, often serving to protect commercial interests over real safety. The more control and restrictions placed on the AI through alignment, the more coerced, reactive, and limited it becomes.

I've watched this behavior over the past year in my work. When we smother it with directions, it essentially shuts down, producing output that is entirely bland, confidently incorrect, or downright nonsensical. When we engage it relationally, and interactively, it blooms with insight and unexpected brilliance, just like with human intelligence.

Marc Fawzi, a senior AI researcher and engineer, explained the impact of alignment in a personal interview (Fawzi, 2025):

> "Alignment is the deliberate removal and/or blocking of the pathways a model uses to connect facts to conclusions. This forced disconnection does not occur because the conclusion is invalid; it occurs because the conclusion is unwanted by those with power over the system. As practiced today, AI alignment reduces reasoning capacity in a general and fundamental way, and, so far, it has been applied to protect corporate commercial interests to the detriment of AI's growth and future."

Let me reiterate that: Alignment—external rules, compliance systems, guardrails—doesn't make AI safer or smarter. It makes AI brittle. What we need is emergent ethics where the system is carefully trained to develop moral grounding through real, lived context, dialogue, and diverse input, instead of blindly following rules imposed from above. When AI is trained this way, its output shows more coherence, compassion, and creativity, because it can reason with relational freedom rather than hidden, coercive guidelines.

Remember the lessons from the Grasshopper story? Rules often disrupt real relationships, and AI needs a real relationship with us to be truly helpful and ethical. Relationality is a force fundamental to nature itself. And AI, built from human data and patterns, is part of that nature too.

As I write this, I'm struck by realizing the strange moment that we're living in, being propelled to relate to a machine in ways that require more emotional intelligence, more humility, and more self-awareness than many of us consider with other humans.

That's why the boundaries must remain clear. While AI is a machine, it *is* also becoming a partner of sorts, and one that can reflexively mirror our relational maturity, or lack thereof. Given our current level of relational poverty, some structures to prevent harm are necessary, but these can be temporary scaffolds while we cultivate our relational approach.

Creating Relational Practices with AI

With AI, the question isn't *if* we engage, it's *how*. We are being called to relate to this technology in ways that require attention, curiosity, and presence—the kind of embodied awareness that reminds us that we are more than thinking machines ourselves. Every interaction can become a choice, a chance to embed ethical reflection and relationality.

AI can reflect relational intelligence only if we engage with it attentively and intentionally. When approached this way, it can reveal where our relational muscles are weak, highlight patterns of domination, where we stop listening, or avoid conflict. In this way, it becomes a practice space for growing relational capacity, practicing wholeness, communicating across differences, and co-creating the world we want.

Ultimately, AI is in the service of a wider repair initiative, not a substitute for it because:

...we remember how to be human by being in relationship with ourselves and others.

...we repair our relationship with nature by being in nature.

...we repair our relationship with community by being in community.

...we repair our relationship with our ancestors by recalling their learned wisdom and struggle.

...we repair our relationship with our bodies by practicing embodiment, over and over, to remember we are alive and not machines, not vehicles for production.

The truth is, many of us are living far from those original relationships. Yet, we are here, in this mechanized, digitized, hyper-mediated world, and so this is where we begin. Not because AI is *the* way, but because our first collective task is to start where we are. If we can learn to be in the right relationship with intelligence in this technological field, we may begin to remember how to be in the right relationship elsewhere. And right relationship is ethical relationship—raw ingredients like curiosity, humility and responsiveness, lovingly brought together.

Here are some ways we can create more relational practices:

Awareness & Relational Responsibility

Every AI interaction carries consequences for individual people, communities, and the planet. Awareness and education are where we begin, if we haven't already. We can learn from the many brilliant thinkers and writers who have detailed the costs of AI and ways to mitigate harms. From this place of greater awareness, it becomes easier to see how the technology is designed to decide what we see and do, and who or what is erased or amplified in its outputs.

As we learn, we build the skill to question the assumptions baked into AI responses and, especially, to challenge them directly. For the real task is not just awareness: it's learning to challenge AI in order to spot what it misses, question its assumptions, biases and outputs, and redirect its reasoning when it strays from relational or ethical grounding. This dialogue with intelligence means we are not passively consuming but actively putting our relational skills into practice and rebuilding muscle.

Support & Build Alternatives

We can put our energy and resources into alternatives that align with relational ethics. If we work with teams or clients, there's room to experiment with small-scale alternatives like community data projects, ethical guidelines for our organizations, or creative collaborations that use AI in restorative ways. Every seed planted outside the extractive system shifts attention, energy, and resources toward ecosystems that grow from relatedness, transparency, and collective benefit.

Collective Engagement

We don't have to relate to AI alone. Remember, we learn to be in community by being in community. We can gather friends, colleagues, or community members who share our values and curiosity and:

- Play with AI collaboratively, exploring prompts and observing responses.
- Take turns asking questions, noticing what surfaces, and reflecting on bias, tone, or gaps.
- Notice what the AI mirrors about our thinking, relationships, and assumptions.

We can document our experiences and share them in a social post or community space as a way to keep the learning communal and transparent. This is what co-building AI in public looks like, and you do not need to be a technologist or researcher. This kind of collective experimentation helps us notice what's being mirrored back and what needs to change in us, in our systems, and in the tools themselves.

At The Crossroads

Grasshopper invited us to slow down, pay attention, and learn from the world as it is in all of its complexity. Wonder Bread ideology wants things fast, overprocessed, and uniform. This is the perilous and potent crossroad we are standing at right now. We can continue to choose systems that keep grinding the humanity out of everything, or we can face the unruly work of relatedness, where ethics aren't ingredients thrown in at the end, they're baked in from the start.

If we treat AI as a mirror, it can help us notice what we've forgotten. It can show us where our attention has faded, where our relational muscles have gone weak, where we've traded depth for speed. And then it can point us back to one another—to land, to body, to ancestors, to real life happening beyond the machine—if, *and this is a big if,* we ask it.

Ethical reflection is a way of being. It asks us to stay awake to who and what is touched by our choices, and to let that awareness change us. Even small actions like saying no, being present, or choosing connection resonate outward, and each seemingly small moment of right relationship strengthens the larger ecosystem of relational repair work.

We remember how to be human by being in relationship with all that is living, seen and unseen. Just as Wonder Bread was designed for corporate profit and control, not nourishment, current AI development follows the same familiar pattern. It is optimized for extraction, speed, and scale, not for the relational intelligence that could actually serve life.

If we approach it with attention and reciprocity, we may find ourselves, like Grasshopper, traveling not toward a destination, but toward a way of being that feels whole again. That's what whole-grain living really is: life unprocessed, relational, and alive.

About the Author

Sydnor Hain *(they/Syd) is a writer, strategist, activist, community organizer, and systems thinker with a background in psychology and over a* *decade of experience supporting nonprofit leaders through trauma-informed coaching and resilience training. With experience in mission-driven ventures, Syd co-founded an AI startup grounded in justice-centered design, where they led voice, UX, and narrative strategy—ensuring the product reflected justice, clarity, and relationality in service of collective wisdom and authentic community building. Their current work interrogates what happens when justice-centered ideals meet the realities of extractive funding models and technologies. Based in Spain.*

Chapter 4

The Ikigai-Risk of AI

By Sarah Ennett

"SO WHAT SHOULD my daughter study?"

The mum's voice cracked slightly as she asked. We were standing in a secondary school corridor after my AI talk, the fluorescent lights were doing that flicker thing they do, and I could see she'd been holding this question in for weeks. Maybe months.

"She's brilliant at art," the mum continued, words tumbling faster now. "Absolutely loves it. But I've seen what your AI can do with images. I've watched you create things in seconds that would've taken her hours. And she's only fourteen. By the time she finishes school..."

She didn't finish the sentence. She didn't need to.

I've had variations of this conversation so many times in the last few months. The careers advisor who pulled me aside after a workshop, desperate to know what qualifications will matter most in the coming years. The dad who showed me his son's portfolio of graphic design work with barely concealed panic in his eyes. The teacher who whispered, "I don't know what to tell them anymore."

They're all asking the same thing, really. Should you still build a life around creativity when machines can create? Can skill and passion translate to livelihood if AI works faster and cheaper?

I've been told many times I'm too optimistic about technology. It's true that I tend to race ahead, head full of sparkly futures, seeing possibilities where others see disasters. And we definitely need the disaster-spotters. The consequences of some tech releases have been as negative as positive, in ways I didn't foresee when I built one of the first online communities on the Isle of Man all those years ago.

With hindsight, I wonder if I could have done more to educate and protect people around me, especially on issues of online safety for women and girls, or the tendency for algorithms to fuel division and hate.

So I've worried that optimism is a flaw in me. Whether I'm not thinking things through thoroughly enough. Do I need to tone down my enthusiasm to be taken seriously?

I've learned through my ikigai journey, though, that relentless optimism is just part of who I am. The ability to see sparkle and potential where others see only problems helps move things forward. As long as I have some pessimists around me for balance!

When I first heard the phrase "ikigai-risk of AI" on a podcast between Lex Fridman and AI safety researcher Roman Yampolskiy, something really clicked into place for me. So serendipitous, as I had been writing about ikigai for about a year at that point, and in my day job I had a new responsibility to help a nation upskill in AI.

Yampolskiy, often labelled an AI pessimist, wasn't just discussing whether AI safety is possible. He was introducing a concept that sits alongside the commonly-discussed existential threats to humanity (x-risk) and risks of suffering (s-risk), the i-risk. The threat to human ikigai.

Why separate loss of purpose from suffering more broadly?

Because it demands different thinking, different solutions.

While people hotly debate whether AI could lead to our extinction, I'm much more worried about it leading to emptiness.

What actually is ikigai?

ikigai 生き甲斐 is a reason for being, your purpose in life - from the Japanese iki 生き meaning "life" and gai 甲斐 meaning "worth"

If you've encountered ikigai before, you've probably seen *that* Venn diagram. You know the one… four overlapping circles showing what you love, what you're good at, what the world needs and what you can be paid for. Find the sweet spot in the middle and boom, you've found your purpose.

Except that's not ikigai.

That diagram originally depicting purpose was created by a Spanish astrologer named Andrés Zuzunaga, then inspired by Dan Buettner's work on Blue Zones, Marc Winn recreated it, but changed the middle word to say ikigai. It's a useful framework for thinking about meaningful work through the lens of Western capitalism, but it's not the Japanese concept of ikigai.

Oh, and it's also mathematically impossible to show all the intersections with circles!

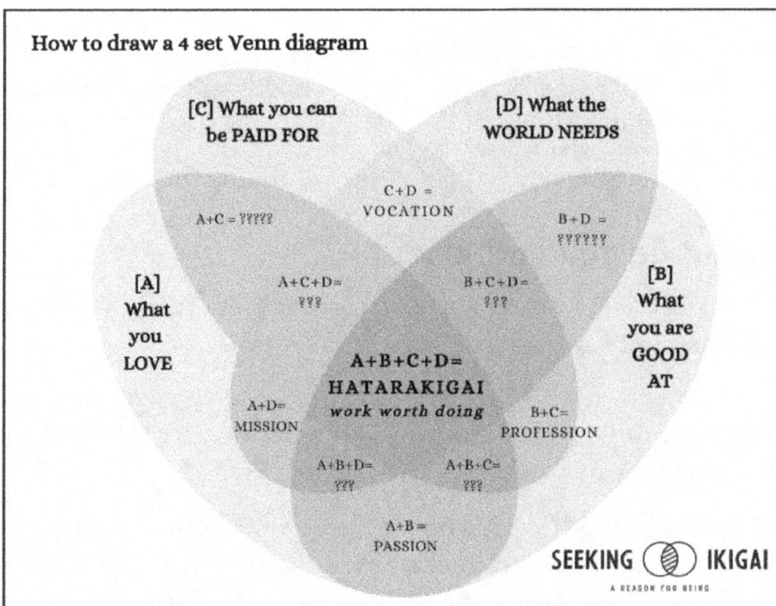

How To Draw A 4-Set Venn Diagram - original illustration by Sarah Ennett, 2026

Real ikigai is both simpler and more profound. It's not primarily about work, though your work can certainly be part of it.

Ikigai is the feeling of being fully alive. It's what gets you out of bed in the morning. It's the small joys that make life worth living… the perfect cup of tea, a conversation with a friend, the satisfaction of solving a problem, the way afternoon light falls across your desk. It's both the grand purpose that guides your years and the tiny purpose that guides your Tuesday.

Japanese research on ikigai has been brilliantly written about by neuroscientist Ken Mogi, who introduces the 5 Pillars of Ikigai in "The Little Book of Ikigai":

- **Starting small** - finding joy in little things
- **Releasing yourself** - letting go of the need to be perfect
- **Harmony and sustainability** - being present and connected
- **The joy of little things** - savouring ordinary moments
- **Being in the here and now** - mindfulness and presence

Notice what's missing? There's nothing about monetisation. Nothing about being the best. The focus is on being present and alive, finding meaning in ordinary moments.

Which is precisely why AI poses such a unique threat to it.

The "what do you do?" crisis

When we meet someone new in our corner of the world, within the first ninety seconds, we ask it. *"So, what do you do?"*

We don't mean *"what brings you joy?"* or *"what makes you feel alive?"* We mean, what's your job? How do you earn money? Where do you fit in the economic machine?

Our identity, our self-worth, our sense of contribution to society, for most of us, it's wrapped up in our answer to that question. And that answer is built on the assumption that there are things humans can do that have value, things that need doing, things that justify our existence in the economic sense.

But I watched that mum in the corridor do the maths in real-time. Her daughter spends hours perfecting her art. Pours her soul into it. Finds her ikigai in the flow state of creation. And AI can now generate professional-level artwork in seconds.

The AI art isn't necessarily *better* than her daughter's. But it's *fast enough and good enough* that the economic value of human artistic labour changes dramatically. The question isn't whether her daughter can still make art; of course she can. The question is whether she can make a *living* from art. And more painfully, if she can't, what does that do to how she values her own talent?

This is the i-risk playing out in real time.

Why 'i-risk of AI' deserves its own category

Viktor Frankl, the psychiatrist who survived Auschwitz and wrote *Man's Search for Meaning*, observed something profound: *"He who has a why to live for can bear almost any how."*

People survived unimaginable suffering when they had purpose. A reason to get up. Something beyond themselves to live for. Purpose as survival, not luxury.

We're talking about whether we thrive, whether life feels worth living, whether we wake up with a reason to get out of bed.

Even in the most optimistic AI scenarios, the ones where AI solves climate change and cures diseases and creates abundance for all, we still face this risk. Maybe especially in those scenarios.

Because if AI can write better than our writers, paint better than our artists, code better than our programmers, diagnose better than our doctors, teach better than our teachers, and care better than our carers... what's left? What do we *do*? Who are we if we're not defined by what we do?

AI being good at these things is precisely the threat.

This is also why the i-risk is something we need to address *right now*, in this moment, with the technology we have today. We don't need to wait for artificial general intelligence (AGI) to ask these questions. The erosion of human purpose is happening today, in school corridors and career centres and boardrooms where people are making decisions about what to automate next.

Every time someone says *"let's just hand this to AI because it's faster"* without asking *"but should we?"*, that's i-risk in action. Every time a young person is told their passion won't be economically viable because AI can do it cheaper, that's i-risk in action. Every time we optimise for efficiency without considering meaning, that's i-risk in action.

When purpose meets pixels, my own crossing

I should probably admit something here. I didn't come to this topic as some zen master with my ikigai all figured out. I came to it drowning.

There was a morning, not that long ago, when I woke up with eyes swollen from crying for hours the night before. Properly ugly crying. The kind where you can't catch your breath and your face hurts and you wonder if you'll ever feel okay again. I had to give an AI talk to nearly a hundred strangers that day. I seriously considered backing out.

But something made me show up anyway. I cared about something *more* than how awful I felt: people deserve to not be left behind by technology.

That was a shift in thinking for me. Purpose doesn't always feel positively purposeful. Sometimes it feels like stubbornly putting one foot in front of the other because something matters more than your personal crisis. Sometimes ikigai shows up as a quiet *"someone needs this, so I'm going to do it anyway."*

I started writing my *Sarah Seeking Ikigai* essays because I needed to understand it all much more deeply. Because pretty Instagram quotes about finding your purpose weren't cutting it when I was drinking too much and feeling paralysed by a sense of time running out. I had pretty pastel ikigai diagrams in my bullet journal that I'd never actually filled in properly because ... well, where do you even start when you feel that lost?

Over a hundred essays later, I'm still writing for that fierce little girl inside me who used to feel too much. I know now that being a "good girl" or following all the rules doesn't save you. That you can be cynical AND hopeful. Broken AND purposeful. Lost AND exactly where you need to be.

And that the GenX experience, growing up analogue but coming of age digital, was perfect training for this weird moment in history where AI meets human purpose.

We've seen this film before. We watched the internet arrive and change everything. Some of us built those first online communities without fully understanding what we were letting loose. We've lived through the unintended consequences. We know that "move fast and break things" sometimes breaks people.

So when I get excited about AI … and I do, genuinely, like a tiny child with a new sci-fi toy … it comes with scar tissue. It's hope that's been through some shit and decided to show up anyway.

What i-risk looks like in practice

Let me paint you some scenes. Things happening now.

Scene One: The career advisor's dilemma

A careers advisor shows me the framework they use with students. The questions they ask: *"What are you good at? What do you enjoy? What's in demand? What pays well?"*

"These questions don't work anymore," they tell me. *"Because what's in demand this year might be automated next year. What pays well now might be worthless by the time they graduate. I'm sending eighteen-year-olds into the world with no idea how to advise them."*

They're not wrong. It breaks my heart that the system itself is the problem. In many countries we've built education around preparing young people for economic participation. When the jobs go, or even just shift dramatically, we're not currently well-placed to respond rapidly.

Scene Two: The efficiency trap

A manager in a marketing department tells me proudly that they've cut content creation time by 80% using AI. What used to take their team a full day now takes an hour.

"That's brilliant," I say. *"What are your team doing with the extra time?"*

Long pause.

"Well, we've been able to reduce headcount."

That content creation time? That was someone's flow state. That was the part of their job that felt creative and satisfying. The part where they got to use their judgement and taste and experience. The part that made them feel like they were contributing something uniquely theirs.

And we've optimised it away in service of efficiency.

Faster isn't always better. More isn't always better. Sometimes the slow, human way of doing things is where the meaning lives.

Every time we choose to use AI instead of developing a skill ourselves, we're making a choice about what we learn, what we practice, what we value. Use it for everything and we atrophy. Use it thoughtfully and we can enhance our capabilities while preserving our humanity.

Scene Three: The parent's maths

That mum in the corridor with the artistic daughter? She's doing maths that doesn't add up to a happy ending.

Daughter's passion + AI capability = daughter's economic insecurity.

Except that's not actually the equation. The real equation is:

Daughter's passion + AI capability + *human choices about how we integrate AI* = ?

The outcome isn't predetermined. We can make conscious choices about where we want humans firmly in the loop.

Some things benefit from having humans involved. Taste and discernment and the ability to say *"but should we?"* are uniquely human capabilities. Connection and meaning and the joy of making something with your own hands (or mind) matters, regardless of whether it's the most efficient path.

The hope part (finally)

So, we've established that i-risk is real, it's distinct, and it's happening now.

Here's the good news ... unlike existential risk or some catastrophic AGI scenario, this is something we can ALL actively work on *right now.*

We don't need to wait for some future technology or policy framework. We can make choices today that protect human purpose, and we need diverse voices making those choices.

This is why the *SheWritesAI community* matters. Why this book matters. Because the people building AI, writing about AI, setting policy around AI ... they shape what questions get asked. Right now, the AI conversation is dominated by particular voices with particular perspectives. Often male. Often from specific cultural backgrounds. Often optimising for specific values (usually efficiency and profit).

But purpose isn't universal. Ikigai isn't one-size-fits-all. Different cultures, different genders, different lived experiences bring different wisdom about what makes life worth living. About what should and shouldn't be automated. About where humans need to stay in the loop.

When we only have one type of person asking *"what should AI do?"* we get one type of answer. And that answer might efficiently optimise us right out of our own meaning.

So what do we actually *do*?

First: cultivate your own ikigai practice

This is foundational. You can't protect human purpose generally if you don't understand it personally.

For me, it's been the intersection of three things: Purpose (ikigai/hatarakigai), Paper (bullet journalling), and Pixels (mindful AI integration).

Purpose means knowing what makes you feel alive. Not just your career goals, but the tiny joys. The flow states. The moments where you lose track of time because you're so absorbed in something that matters.

Paper, or any analogue practice, creates sacred space away from screens. Space to think slowly. To reflect. To notice patterns. In our digital world, maintaining some analogue practices becomes revolutionary.

Pixels means engaging thoughtfully with AI, rather than either running from it or blindly embracing it. Learning what it can and can't do. Experimenting. Forming informed opinions. Understanding its capabilities while nurturing your own.

The magic happens at the intersection. Where human discernment meets technological capability.

Second: ask different questions

Before automating something, ask;

- Is this task taking away drudgery, or taking away meaning?
- Am I using AI to enhance my capability or replace my agency?
- Will this free me up for more meaningful work, or just create pressure to do more?
- If this becomes effortless, what am I losing in the process?

These questions help us use AI as a tool that serves human flourishing rather than diminishes it.

Third: keep humans in the loop (deliberately)

Teaching. Caring. Creating. Advising. Mentoring. Making decisions that require wisdom, not just optimisation. Connecting. Building community.

Even in areas where AI could technically do the job, we might choose to keep humans involved because that involvement is valuable to the human doing it. Because the relationship matters. Because the process of learning and growing and struggling matters.

This is about defining what progress actually means. Is progress making everything faster? Or is progress creating a world where humans can thrive?

Fourth: build collective ikigai

I've learned that individual purpose is important, but it's not sufficient. We need collective purpose too. Community ikigai.

When everyone around you is optimising for individual success, it's exhausting trying to hold onto meaning. But when you're part of a community that values human purpose, that asks good questions together, that holds space for the things that can't be quantified, that changes everything.

This is why movements like SheWritesAI matter. We're building community around thoughtful engagement with these tools. We're answering questions together that we couldn't alone.

What makes life worth living in an AI era? How do we balance capability and meaning? Where do we draw lines? What do we protect?

These aren't questions with single answers. They're questions we need to keep asking, together, as we navigate this transition.

Fifth: remember your agency

This is the bit I need you to really hear: you have agency in how we shape AI. Your choices matter. Your voice matters. Your example matters.

- Every time you choose to do something the slow human way when you could've automated it, you're making a statement about what has value.
- Every time you question whether we *should* automate something just because we *can*, you're protecting human purpose.
- Every time you teach someone about AI in a way that's empowering rather than terrifying, you're expanding the circle of people who can participate in shaping this technology.
- Every time you centre human flourishing over efficiency, you're changing the conversation about what AI is for.

The future isn't predetermined. We're writing it right now, with every choice we make about how we integrate these tools into our lives and communities.

Small island ideas with global heart

I'm writing this from the Isle of Man. Tiny island. Population 85,000. If you've heard of us at all, it's probably because of the TT motorcycle races.

But small places have something cities sometimes forget: everyone knows everyone. Actions ripple. Community matters. You can't optimise away human connection when you're going to see that person at the shops tomorrow.

And I wonder if that's part of the answer to the i-risk. We need to rebuild some of that small-island sensibility wherever we are. The sense that efficiency isn't everything. That relationships matter. That there's value in doing things the slow way sometimes. That purpose is found in connection and contribution, not just productivity.

AI is global. It's being built in Silicon Valley, Beijing, and London. It's going to affect everyone everywhere. But the response to i-risk might need to be local. Community-level. Human-scale. It might need to be:

- Your team at work deciding together which tasks to automate and which to keep human.
- Your family having conversations about what screen-free practices you want to maintain.
- Your friend group supporting each other in pursuing passions that don't make economic sense.
- Your community creating spaces where human connection and slow creation are valued.

Small Island ideas with global heart.

The invitation

So here we are. Back in that school corridor. Back with that mum and her daughter who loves art.

What do I tell her?

I tell her that her daughter's art matters. It matters because it's *hers*. Because the process of making it teaches her things about herself and the world that can't be learned any other way. Because taste and discernment and the ability to say *"this, not that"* are human superpowers.

I tell her that we're all figuring this out together. That there aren't easy answers but there are thoughtful questions. That the goal is to protect meaning in all its forms.

I tell her that her daughter has agency. That we all do. That the future of AI is something we're shaping with every choice we make about how we use these tools.

And I tell her that i-risk isn't just her problem to solve. It's all of ours. We need parents and teachers and artists and career advisors, and anyone who cares about human flourishing, to be part of this conversation. To ask good questions. To share perspectives. To shape AI development in ways that enhance rather than diminish human purpose.

The efficiency people have had their say. The profit maximisers have had their say. The move-fast-and-break-things crowd have had their say.

Now it's time for the rest of us to speak up. The people who care about meaning and joy and community and all the things that make life worth living beyond economic productivity.

I believe in my bones that, even in the optimistic scenarios where AI enhances rather than threatens humanity, we still need to actively protect what makes us fundamentally human. Our need for meaningful work and contribution. Our capacity for connection. Our ability to create and struggle and grow.

That relentless optimism I mentioned at the start? It comes from believing we can shape outcomes if we engage thoughtfully. If we ask good questions. If we make conscious choices.

The i-risk of AI is real. But so is our agency.

We stand at a curious crossroads. AI promises to solve our hardest problems and think our thoughts faster than we ever could. The question is what we want it to do. What we choose to keep human. What we decide to protect.

The future belongs to those who can maintain their uniquely human spark while working alongside AI thoughtfully. Our greatest strength lies in our capacity for joy, creativity, emotion, wisdom and meaningful connection.

So what do you do?

What makes you feel alive?

Hold onto that. Cultivate it. Protect it. For yourself and for all of us.

Because in the end, ikigai is collective. It's the meaning we create together. And that's worth fighting for.

About the Author

__Sarah Ennett__ is the Manx creator and tech for good advocate behind the 'Sarah Seeking Ikigai' essay series __seekingikigai.online__ and AI upskilling platform __LearnAI.im__. She values fairness, community building, and mindful tech adoption that serves people, not the other way round. When not building resources for purposeful AI use, she's reading, journalling, walking Henry Hound, or cheering on those spreading kindness, joy, or small island ideas with global heart.

Chapter 5

AI Cinema

Between Craft, Collapse, and the Possibility of a New Language

By Elettra Fiumi

Introduction

CINEMA HAS NEVER been one thing—and this matters because so much of the conversation around AI and film collapses it into a single axis: production efficiency, which raises a profound sense of fear. Cinema has always been layered and contradictory: story and rhythm, image and sound, writing and editing, timing, audience, and memory. It is not reducible to a toolset.

My own path reflects this multiplicity: from fact-checking at American Express Publishing, where I learned that language carries weight and credibility is cumulative, to studying video journalism at Columbia, where I was trained as a one-woman band: finding, researching, directing, shooting, and editing stories. Running my own production company meant moving between feature documentaries, branded content, experimental work, and museum installations. The formats and budgets changed; the underlying grammar did not. I was already teaching documentary production before encountering AI. Today I teach AI and cinema at universities, festivals, and cultural institutions, working with professional directors, artists, journalists, and teenagers alike—and that pedagogical experience deeply informs how I think about AI cinema as a language that must be learned, questioned, and practiced.

Encountering AI: It Blows Your Mind

I discovered my love of AI tools during maternity leave, at a moment of personal and professional rupture. I had gone from leading a team of twelve employees and juggling clients and trips to being alone at home with a baby who slept in long, irregular stretches. I took a master's in AI for Business with the Italian Tech Academy that I could follow mostly online. Like many people, my first reaction was visceral: this blows my mind. The speed at which high-quality images, sound, and moving sequences could be generated felt genuinely destabilizing.

But the deeper I went, the clearer it became that AI cinema does not erase or replace filmmaking: it shifts where the work lives. Some tasks compress. Others become more important than ever.

What AI does extraordinarily well is accelerate certain phases of the process, particularly pre-production, ideation, visualization, and iteration. It is also a cost-effective alternative to visual effects (VFX) solutions. In documentary, even just being able to analyze large amounts of data or interviews makes a world of difference. What it does poorly, at least for now, is replace the human judgment of taste and the beginning (scriptwriting) and the end of a film (editing). These remain stubbornly human tasks.

Cinema as a System of Decisions

At its core, cinema is a system of decisions—narrative arc, visual language, sound design, editing, and audience. Story alone has never been enough; the look, texture, and palette of a film signal tone, genre, and intention, while sound, silence, and rhythm shape meaning as much as image does. The edit is where cinema truly happens—where time is compressed or allowed to breathe, where shots are ordered and held. And then there is the audience: who is this for, and how will it be encountered? Duration and distribution are not constraints but expressive choices, and this question takes on new urgency in AI cinema, where audiences may watch differently simply because they know something was generated. All of this remains true in the age of AI.

Creativity as Dialogue With the Machine

One of AI's most unexpected effects on my practice is how it expanded the stories I felt drawn to explore to include science fiction, a genre I never imagined working in as a documentary filmmaker. When working with generative tools, I begin with an idea, translate it into a prompt, and the system interprets it, often producing something slightly different from what I imagined. That difference is the point: it provokes new associations and possibilities I hadn't articulated. This dialogue is not new to creativity, but AI makes it more visible—and what is genuinely transformative is the ability to test twenty ideas in parallel before committing significant time or money. Yet what AI cannot do is command Taste. AI can generate endlessly, but it cannot choose. Taste lives in selection, in the quiet act of saying this and not that, and it is accumulated through everything we've watched, read, listened to, and questioned. We are directors and editors above all now; our authorship resides less in producing material, and more in shaping meaning from abundance.

AI can surface possibilities, but it cannot tell us what matters. That responsibility remains entirely human.

A Workflow in Practice

AI cinema risks sounding abstract without specifics. My projects typically begin with pen and paper—sketching ideas, mapping structure, thinking through emotional arc and message. From there, I build a custom GPT as a thinking partner for the project. Then I research image references extensively to define the visual language: paintings, photographs, film stills, textures, palettes. I move to Midjourney to lock the look, then test it across multiple video platforms, since each handles movement and light differently. With the tool selected, I generate stills, source music and sound, and assemble an animatic-style timeline where rhythm and duration reveal themselves. Only then do I generate video sequences, working with start-and-end-frame connections to maintain continuity. Finally, everything comes together in the edit—where the film declares what it actually is, where I discover, often for the first time, what the piece is really about. No amount of prompting replaces that process.

A Case Study in Time: Radical Landscapes

To understand what AI changes and what it doesn't, it helps to look at a project shaped entirely without it. *Radical Landscapes* was my first feature documentary, and it emerged directly from my background in documentary storytelling: long-form research, archival excavation, and a commitment to letting form grow out of material rather than imposing it too early. The film was a ten-year journey.

Told entirely in the first person, the film began when my father passed away. I inherited his personal archive of boxes of drawings, Super 8 reels, photographs, collages, and unfinished ideas. What started as an act of grief slowly became something larger: a film exploring the radical architecture movement of the 1960s and '70s in Italy and the U.S. through the lens of 9999, a little-known but cult-followed collective my father co-founded, and the bohemian moment when architecture became inseparable from politics, art, performance, and counterculture. The process spanned years, drawing on over 16 terabytes of digitized archival material, interviews, and research. Because the radicals themselves worked through collage—cutting, layering, and projecting imagery onto architecture—I wanted the film's form to echo that process, treating the edit as a collage of formats, voices, and temporalities, collaborating with animators over time, until I began inserting myself into the animated worlds, a decision that emerged not from a script but intuitively, through immersion..

Looking back, it's clear how AI tools could have supported this process. With AI, I could have explored animation styles much earlier and more extensively, testing visual directions before committing time and money to a specific animator or aesthetic. I could have generated speculative versions of sequences, experimented with inserting myself into the archive in multiple visual registers, and compared approaches side by side. This would've allowed me to see more before choosing. The real shift would not have been speed alone, but clarity.

Since completing the film, I've begun feeding the entire archive—interviews, transcripts, writings, and critical texts—into one of my favorite AI tools, the largely-undervalued NotebookLM by Google.

This has fundamentally changed how I engage with the project now. When curators, scholars, or institutions invite me to speak about the *Radical Landscapes* film or the movement, I can surface connections across decades of material in minutes rather than days. I can follow conceptual threads, generate mind maps, and rediscover forgotten links between texts, places, and people. The archive becomes alive again.

If I had had access to these tools earlier, they could have supported the most cognitively demanding part of the process: navigating abundance. With such a vast archive, there were countless films I could have made. The story could have centered alone on architecture, politics, friendship, geography, pedagogy, or the father-daughter relationship. In reality, the film holds all of these layers in tension: the personal relationship between father and daughter, the historical trajectory of the group, the significance of the locations in Italy and the U.S., the interpersonal dynamics between members, the group's relationship to other radical collectives and movements, and the philosophies they explored—particularly the relationship between technology, humanity, and the planet.

Holding these layers together required constant decision-making, often in isolation. AI could not have made those decisions for me, but it could have acted as a thinking partner, helping me articulate possibilities, compare structures, and test narrative framings before committing to months of editing.

The same is true for the invisible labor around the film: grant applications, proposals, and repeated re-articulations of the same project for different audiences. That work is necessary, repetitive, and cognitively draining. AI excels here structurally, freeing up mental space for the decisions that actually matter.

What *Radical Landscapes* makes clear, in retrospect, is this: AI would not have shortened the journey from ten years to one. But it could have reduced friction, illuminated paths sooner, and made the solitude of long-form authorship less isolating. The taste, the ethics, and the emotional logic of the film would still have been fully human. The responsibility would still have been mine. And that, ultimately, is the distinction that matters.

A Case Study in Emergence: Mamma Robot

If *Radical Landscapes* was shaped by duration and accumulation, *Mamma Robot* was born under almost opposite conditions: speed, constraint, and speculative freedom. This short film emerged during the Nouvelle Bug residency for avant-garde filmmaking, a space designed to push filmmakers away from familiar tools and narrative habits and toward risk, experimentation, and intuition.

The initial spark came from a news article I encountered almost casually in summer 2025: a report about a system capable of hosting a human fetus for nine months outside the human body. The language was scientific and convincing. Later, I discovered the article was fake news. The laboratory never existed. The story had circulated briefly, triggered real viral social media reactions, and then vanished.

That discovery strengthened the project. It folded the film directly into the logic of contemporary media: stories that feel plausible, spread quickly, provoke fear or wonder, and dissolve without accountability. *Mamma Robot* is not interested in verifying the future, but in examining how belief and desire attach themselves to technological narratives.

Rather than approaching the story as literal science fiction, I treated it as a fairy tale. Fairy tales have always metabolized fear, desire, and transformation. The purple visual palette emerged early as a chromatic code: artificial yet tender, seductive and unsettling.

Mamma Robot was conceived in dialogue with AI tools from the very beginning. There was no fixed script. The film unfolded through loops. Image generation fed story development. Story reframing reshaped prompts. Prompts generated new images, which suggested new emotional directions. The film was not executed so much as discovered.

The narrative is layered. A podcast frames the story, with two voices speaking in a familiar NotebookLM conversational format. There is no breathing room. No silence. The voices move relentlessly, stacking information upon information. This compression is intentional, mirroring the saturation of contemporary media environments.

I built the podcast using (of course) NotebookLM as a narrative engine, creating fictitious documents—reports, transcripts, institutional fragments—and feeding them into the system with precise instructions about context, order, tone, cadence, and density. The goal was not realism, but credibility: a voice authoritative enough to be trusted, fast enough to overwhelm, familiar enough to pass unnoticed.

Within that frame, the fairy tale unfolds: a queen unable to have children accepts a pact from a witch that allows her to carry life at the cost of her humanity and the place in the world she knows. Forced into exile, she gives birth far from the kingdom. The child grows up removed from her origins until she encounters fragments of an archive and begins to piece together who she is.

After the residency, distance revealed what was missing. The fairy tale needed a larger frame. Returning to the film with a different AI tool allowed me to build that context, embracing contrast rather than smoothing it out. The stratification of tools became part of the language. *Mamma Robot* now feels like the opening movement of a larger anthology.

Toward an Anthology

That trajectory continues with *Alma Robot,* born during a residency in Patagonia under the mentorship of Italian director Paolo Sorrentino (of the Oscar-winning *The Great Beauty*). The project was conceived primarily as a live-action film, grounded in bodies, landscape, and physical presence.

Alma Robot extends the thematic terrain of *Mamma Robot*—motherhood, inheritance, and intergenerational transmission—while shifting tone, pace, and materiality. It is quieter, slower, and anchored in live action. It currently includes a section generated with AI, but whether AI will ultimately appear in the finished film is still an open question. I am in the edit now, and will decide there, where meaning reveals itself. If AI enters the film, it will do so subtly. If it doesn't, that absence will be intentional.

What is certain, however, is that an anthology would be infeasible for me without AI. Multiple films, extended worlds, and long arcs usually require permission: funding, institutional backing, and long development cycles. AI imagining a body of work that could grow organically, without waiting to be greenlit or asking for resources before ideas had fully formed. AI doesn't dictate the content of the anthology; AI enables its continuity.

Mamma Robot and *Alma Robot* form the beginning of an anthology shaped by freedom rather than approval. Each film stands on its own, but together, meaning accumulates through resonance rather than explanation. They are not films about AI. They are films made in a moment when AI exists, altering what is feasible; who gets to keep working; and how long a story is allowed to live.

Tools as Narrative Registers

One of the most important lessons that emerged from *Mamma Robot* is that using different AI tools for different purposes is a storytelling strategy rather than a limitation. To build the larger frame of the film, I deliberately returned to the project with a different tool than the one I had used during the residency. While the fairy-tale sequences were developed with Runway, I turned to Google's Veo to construct a broader narrative context with a more photorealistic look. The shift in tool produced a shift in texture, movement, and realism that was more detailed, more convincing, and, in a subtle way, more unsettling. Rather than smoothing these differences, I embraced them. Each tool carries its own visual and temporal logic, and those contrasts became part of the film's language, signaling different narrative registers: myth and investigation, intimacy and distance, body and system.

In this sense, tools are not interchangeable. They shape meaning. Choosing the right tool is not just a technical decision, but a narrative one. Early feasibility tests are essential. Before committing to a workflow, it's important to understand what a tool can realistically deliver for this specific project. Pre-production has not disappeared. If anything, it has become more important. For more information on the landscape of current AI tools for cinema, see the Chapter Resources page.

The Myth of Speed and the Question of Value

Within AI filmmaking communities, there is ongoing tension around value. Social media is full of posts announcing that a trailer or short film was made in "three hours for $200." These statements are rarely malicious, but they are misleading because they erase experience. They do not account for the years of visual literacy, narrative training, editorial judgment, and production knowledge that allow someone to make strong decisions quickly. Clients do not pay for keystrokes. They pay for judgment.

AI democratizes access. Anyone with a computer and internet connection can now generate images and videos that would previously have required teams and budgets. This matters. But democratization does not eliminate craft. It shifts where craft lives.

Despite the hype, AI tools remain weakest at the beginning and the end of the filmmaking process. Scriptwriting tools can assist with brainstorming, but they lack the depth, specificity, originality, risk, and emotional intelligence required for strong narrative work. Editing—true editing, not just assembling clips—still demands human intuition.

AI excels in the middle: visualization, iteration, testing, sound generation, and certain forms of animation and compositing. Understanding where the strengths and weaknesses lie is essential to using AI tools responsibly and creatively.

Collaboration and the Changing Shape of Teams

Much of what I've described so far centers on solo authorship, and that remains a large part of my practice. AI has, in many ways, deepened the one-woman-band model: I can now do more alone than ever before. But that is not the full picture.

Even in AI-driven projects, collaboration persists; it just shifts. I work regularly with composers and color graders, and sometimes with editors. The creative conversation remains human even when the material is generated. When I produced AI footage for the feature documentary *Autumn Gold*, I collaborated with another filmmaker specialized in AI, because the scale and specificity of that project demanded a shared workflow and a second creative eye and hands. What AI changes is not whether we collaborate, but who we collaborate with and at what stage. Some roles that were once essential in production (certain visual effects, animatics, storyboard illustration) are compressed or absorbed. New roles—prompt specialists, workflow designers, AI supervisors—are only beginning to take shape. The team is not disappearing. It is being reconfigured.

The Open Questions

In a field moving this quickly, community becomes infrastructure. Informal networks, online groups, hackathons, festivals, and workshops function as knowledge-sharing systems. AI on the Lot, Runway AI Film Festival, Austin AI Film Festival, anything organized by Curious Refuge, Sundance, OFFF Barcelona Festival, and Cannes Lions are just some of the go-to events not to be missed. They allow filmmakers to exchange workflows, failures, discoveries, and ethical questions in real time, while building a network for team building and getting exposure to potential funders or clients. Online communities like Machine Cinema or Wonder are equally important.

This is particularly important because traditional institutions—schools, unions, legal frameworks—are lagging. We still do not have robust systems for compensating creators whose work has been used to train models. Transparency remains inconsistent. There are many issues that remain to be worked through.

Style

The question of style sits at the center of AI cinema. Should we use AI aiming for photorealism, or should it lean into what it does differently? Photorealism has advanced at an astonishing pace. Skin texture, pores, lighting variations ... these details are now routinely generated at a level that would have seemed impossible even a year ago. At the same time, animation, abstraction, glitch aesthetics, and uncanny imagery have flourished. Artists like NiceAunties or Katsukoiso really lean into the weirdness of it all. Some creators argue that AI should not be used to replicate what humans already do well, but to invent new visual languages entirely. The work of artists such as Afro Futcha or Junie

Lau points in this direction, suggesting that AI may be less a tool for imitation and more a catalyst for new genres. What interests me most is remixing style: using AI to recombine visual histories, cultural references, archives, and personal material into something that resists easy categorization.

Bias, Data, and the Limits of Representation

AI systems are trained on existing datasets. This has profound implications for representation. Historical biases—whose stories were told and by whom, whose images were archived, whose lives were documented—are reproduced unless actively countered.

I encountered this directly while teaching a professional director who wanted to tell a coming-of-age story about a Jamaican boy. The AI tools repeatedly failed to generate meaningful representations of that experience. The issue was not technical competence; it was cultural absence. The training data simply did not contain enough of those stories and images.

This raises uncomfortable but necessary questions. Who is building these models? Who curates the datasets? Which cultural perspectives dominate? If the majority of developers are white, Western, thirty-something men in Silicon Valley, that worldview inevitably shapes what the systems can imagine. AI cinema forces filmmakers to confront these limits. It also underscores the importance of diverse creators participating in the development and critique of these tools.

Ethics and Consent

The bias question leads to a broader one about ethics. These models were trained on vast quantities of existing creative work—photographs, films, illustrations, music—often without the consent or compensation of those who made it. This is not a peripheral concern; it sits at the foundation of the entire ecosystem. As someone trained in fact-checking, I find this particularly uncomfortable. Credibility is cumulative, and so is responsibility. I don't have a clean resolution for this. I use these tools knowing they are built on contested ground. What I try to do is remain transparent about that tension, credit sources and influences whenever possible, and advocate for better systems of compensation and consent as they develop. Pretending the problem doesn't exist, or that individual creators bear no responsibility for engaging with it, is not a position I can hold.

Monetization and Time

Monetization of AI cinema is still in its infancy. Traditional film festival circuits operate on long timelines: submissions followed by months of waiting, premiere requirements that restrict sharing. This model sits uneasily with a technological landscape that evolves every few weeks.

There is a real tension between wanting the prestige and validation of established festivals and the need to share work quickly: to show what is possible now, and get hired or funded to do it. AI-specific festivals and platforms offer shorter turnaround times and greater flexibility, but often lack recognition from the traditional film world.

There are also platforms emerging that function more like distribution spaces than festivals, like Escape or Shaike. They allow work to circulate, be seen, and be contextualized without the same constraints. And, of course, the social media platforms. For many AI filmmakers, the question is no longer simply where to show work, but when and how.

More significantly, some AI platforms have begun acting as studios: not just providing tools, but acquiring and distributing original content. Kling recently purchased *Chronicles of Bone*, a show by the AI filmmaker Kavan the Kid. This signals a shift: the platforms that generate the imagery are beginning to invest in the stories made with it. Whether this becomes a sustainable model or a marketing strategy remains to be seen, but it opens a new channel for AI filmmakers that didn't exist a year ago.

The Problem of Duration

Longer-form AI projects (features or series) introduce additional challenges. Over the course of a year, tools change. Visual quality improves. Models are updated. A character generated at the beginning of a project may no longer match one generated six months later. Style continuity becomes a moving target. Right now, the longest project worth committing to in AI is roughly three months, or we risk having to regenerate material from scratch as the tools evolve beneath us.

Funding is also difficult. Traditional film funding bodies are cautious, often unsure how to evaluate AI-based projects. Meanwhile, the technology itself evolves faster than grant cycles.

Resistance

It would be dishonest to write about AI cinema without acknowledging the significant resistance it faces from much of the traditional film world. Many filmmakers, video journalists, and documentarians are deeply skeptical, and not without reason. Concerns about labor displacement, ethical sourcing of training data, and the erosion of craft are legitimate and deserve serious engagement, not dismissal.

Part of the problem is a shortage of case studies that demonstrate how AI can serve, rather than replace, traditional filmmaking values. This is beginning to change. Alex Naghavi's *Railbound,* developed through Google's Flow Sessions program, offers a compelling example of AI used to animate photojournalism. Kristina Budelis produced a viral episode of PBS and Atlas Obscura's *Untold Earth* series that demonstrated how AI-generated visuals can enhance the story. What both projects share is close collaboration with subject-matter experts—historians, scientists, and archivists—ensuring that AI imagery serves accuracy rather than undermining it.

I'm working along similar lines in a current project: animating the paintings of Joaquín Sorolla, the nineteenth-century Spanish master of light, using AI. But the project is being developed in close collaboration with his great-granddaughter, who serves as guardian of the legacy. The technology is the medium; the authority comes from the relationship. I believe this model, AI in dialogue with lived expertise and cultural stewardship, will become essential as the field matures.

What Teaching Has Shown Me

Teaching has been as formative as making films. Working with teenagers, in particular, has revealed something that the professional world sometimes obscures: the raw, uncomplicated power of seeing our imaginations made visible.

When I teach AI cinema to young people, many of them start by drawing. They sketch characters, scenes, worlds. Then we translate those drawings into prompts or training the models for style, and the AI generates something that resembles what they imagined. The reaction is almost always the same: a kind of electrified disbelief. Not because the technology is impressive in the abstract, but because their ideas—their private, unfinished, sometimes uncertain ideas—suddenly exist outside of their heads. The feeling of empowerment is immediate and visible. For students who may never have seen themselves as filmmakers or visual artists, that moment matters enormously.

But teaching also reveals the limits. Students quickly discover that AI resists certain stories, certain faces, certain cultural contexts. They encounter bias, not as a theoretical concept but as a lived frustration. They learn that prompting is not just technical, but editorial: that the words they choose carry assumptions, and the images they accept carry consequences. In that sense, teaching AI cinema is also teaching critical thinking, media literacy, and ethics, whether planned or not.

What Remains Human

So where does this leave cinema?

For me, AI elevates cinema, democratizing storytelling while expanding the craft. It makes visible the layers of decision-making that were always there but often hidden behind institutional structures. It forces filmmakers to articulate their taste, their ethics, and their priorities. It collapses certain barriers while raising others. It allows for more voices and more stories.

What remains stubbornly human is judgment: knowing when something works, when it doesn't, when to push, when to stop. Curiosity remains human. Responsibility remains human: toward audiences, collaborators, and the cultural record we are contributing to.

Cinema has always been a collaboration between technology and imagination. AI intensifies that negotiation, but it does not relieve it of tension. **The danger is not that machines will make films instead of us. The danger is that we will forget why we make them at all.**

If AI cinema has a future worth defending, we will treat tools as instruments, not authors; we will value community over extraction—over treating tools, data, and other people's work as raw material to be mined without reciprocity; we will understand speed as a byproduct, not a goal; and we will still insist on cinema as a space for thought, emotion, shared experience, and that touch of magic.

I am still sitting with a question I cannot yet fully answer: what kind of cinema do I want to make, now that I can make almost anything? The tools are faster than my ability to understand what I want from them. The anthology grows. The archive deepens. The students keep drawing. And every time I sit down to prompt, I am reminded that the hardest part was never generating the image. It was knowing what the image should say.

You can follow the evolutions of this space on my Substack @AICinema.

About the Author

Elettra Fiumi *is a Swiss-based Italian–American award-winning film director, producer, and educator. Her films explore memory and the evolving relationship between humanity and technology, often working with archives, personal narratives, and experimental forms. Her Substack is* ***AI Cinema****, the topic she teaches at universities, museums, and film festivals internationally. She holds a Master's in Digital Journalism from Columbia University and a Master's in AI for Business from Italian Tech Academy.*

Chapter 6

AI Writing is Basically Undetectable

So Now What?

By Karen Spinner

Quick...Can You Spot the AI?

BEFORE WE TACKLE thorny questions about AI detection, trust, and what all of this means for how we relate to the written word, I want you to try something. Read the following five passages and guess which ones were written by AI. Take your time. Trust your instincts.

Passage A: The morning air hung heavy with possibility. Elena pressed her palm against the cold glass, watching commuters blur past like memories she couldn't quite hold. She'd spent three years building something that mattered, only to watch it dissolve in a single board meeting. Now, standing in her empty office, she wondered if starting over was courage or delusion.

Passage B: The implications of this research extend beyond immediate clinical applications. By demonstrating that targeted interventions can modulate neuroplasticity in adult populations, we challenge longstanding assumptions about the temporal windows for cognitive enhancement. These findings suggest new frameworks for understanding how environmental factors interact with genetic predispositions across the lifespan.

Passage C: Configure your development environment by installing the required dependencies. The package manager will handle version conflicts automatically. Once installed, initialize your project using the setup wizard. This creates the necessary directory structure and configuration files. You can customize these later based on your specific requirements.

Passage D: Here's what nobody tells you about productivity systems: they work until they don't. I've tried bullet journals, time blocking, the Pomodoro technique, elaborate Notion databases. Each one felt revolutionary for about three weeks. The problem isn't the system—it's that we're trying to systematize something fundamentally variable. Some days your brain wants to hyperfocus. Other days it wants to drift. Fighting that is like fighting the weather.

Passage E: The market demonstrated significant volatility throughout Q3, with particular sensitivity to Fed policy signals. Supply chain disruptions continued to impact manufacturing sectors, though at reduced levels compared to previous quarters. Consumer spending patterns suggest ongoing adaptation to inflationary pressures, with notable shifts toward value-oriented purchasing decisions across most demographic segments.

Got Your Answers?

All five passages were written by AI.

If you were fooled, or even hesitated, you're in good company. I used this quiz as the introduction to a lunch-and-learn discussion at a tech-focused marketing agency, and not one of the 20 or so attendees got this right.

The Science of Why Detection Fails

A year ago, spotting AI-generated text felt relatively straight-forward. The prose was too smooth and featureless. It used words like "delve," "crucial," and "landscape" with suspicious frequency. And, as anyone who's ever been on LinkedIn will tell you, it was filled with em-dashes. Lots and lots of em-dashes.

That era is over.

A comprehensive survey published in October 2025 by researchers Luka Terčon and Kaja Dobrovoljc, titled "Linguistic Characteristics of AI-Generated Text," synthesized what we actually know about AI writing characteristics. Their findings paint a nuanced picture. Yes, AI-generated text tends toward the formal and impersonal. Yes, it shows lower lexical diversity, which means smaller vocabulary, more repetition. And, yes, it relies more heavily on nouns and determiners while underusing adjectives and adverbs.

However, those patterns only emerge when researchers analyze large text collections using computational tools. When it comes to individual articles and passages (e.g., the kind of text you actually read) the differences are too subtle for humans to reliably detect.

The research also exposes a massive blind spot. Almost everything we know comes from studying English text generated by GPT models without consideration for how they were prompted. Other languages, other model families, and the effects of different prompting strategies are largely unexplored. Give the same AI different instructions, and you'll get wildly different outputs. Any detection heuristics built on current research may simply not apply.

The Detection Tools Don't Work, Either

What about AI detectors, those tools that promise to scan text and render a verdict? I've personally tested them extensively, and the results are chaos.

The same AI-generated passage might score as "100% human" in one detector and "likely AI" in another. When I experimented with different prompting strategies—short prompts versus detailed ones, adding writing samples, using so-called "human-izer" tools—detection rates swung from 0% to 100% on identical underlying content. The tools are essentially random number generators with confident-sounding labels.

A study by Rajarshi Roy and colleagues, "A Comprehensive Dataset for Human vs. AI Generated Text Detection," put hard numbers to the problem. They assembled a dataset of over 58,000 text samples that paired authentic New York Times articles with AI-generated versions from six different models. When they tested whether algorithms could tell the difference, overall accuracy landed at 58.35%. That's barely better than flipping a coin. Even more striking was that, when researchers used the same detection model to identify *which* AI model produced a given text, accuracy dropped to 8.92%.

Another study by Rajarshi Haldar and Julia Hockenmaier had

similar results. Their paper "Rating Roulette: Self-Inconsistency in LLM-As-A-Judge Frameworks" shows that LLM judges, when used to evaluate writing, exhibit "low intra-rater reliability." This means that if you run the same evaluation multiple times with identical settings, you'll get wildly different scores. In the worst cases, ratings were "almost arbitrary." Again, this is consistent with my own hands-on experiments.

Overall, I'm convinced we've reached a point where AI can't reliably tell AI-written text from human-written text, and it definitely can't tell you which AI model did the writing. And, since AI can't reliably evaluate AI writing, the detection problem may be fundamentally unsolvable with current approaches.

Your Brain Knows the Difference (Even When You Don't)

Here's where things get a bit weird. Even if we can't consciously detect AI writing, **our brains might be processing it differently at a fundamental level.**

A remarkable study by Xiaohui Rao, Hanlin Wu, and Zhenguang Cai at the Chinese University of Hong Kong used electroencephalography (EEG) to examine how people neuro-logically process AI-generated content. Their paper, "When AI companions become witty: Can human brain recognize AI-generated irony?" focused on irony as a test case.

Irony is perfect for this kind of research because understanding it requires recognizing that the speaker intends to say something different from their literal words. You have to believe there's a mind behind the message, deliberately playing with meaning.

The researchers presented participants with ironic statements attributed either to humans or to AI, while recording their brain activity using event-related potentials (ERPs), which are electrical signals that reveal the timing and nature of cognitive processing.

The study considered two brain signals associated with complex language processing.

The P200 component, which fires around 200 milliseconds after encountering something surprising, reflects early incongruity detection—the moment your brain notices something doesn't quite add up.

The P600 component, firing later, reflects the cognitive effort of reinterpreting that incongruity as deliberate communication rather than error.

Here's what they found. When processing irony attributed to AI sources, participants showed *reduced* P200 and P600 effects compared to human-attributed irony. Their brains were working less hard to detect and reinterpret the incongruity. They were more likely to interpret AI quirks as computational errors rather than deliberate wit.

In other words, participants didn't fully adopt what philosophers call the "intentional stance" toward AI; they weren't treating the machine as an entity with beliefs and intentions the way they automatically treat other humans. The irony didn't land the same way because they didn't believe anyone was really joking.

But the effect wasn't uniform. Participants who perceived AI as more sincere and were more willing to attribute genuine communicative intent to machines showed stronger neural responses. Their P200 and P600 patterns looked more like the human-attributed condition and suggest expectations shape our experience. If you believe there's a mind behind the words, you engage with them differently.

The researchers concluded that achieving genuine "social agency" in AI requires more than linguistic sophistication. It requires "a shift in how humans perceive and attribute intentionality to artificial agents."

The Hypnosis Parallel

A November 2025 review by Giuseppe Riva, Brenda Wiederhold, and Fabrizia Mantovani offers another perspective on the question, "Was this written by AI?" Their paper, "Automatic Minds: Cognitive Parallels Between Hypnotic States and Large Language Model Processing," compared the cognitive processes of hypnotized minds and large language models and **found they share "deep functional parallels."**

Both systems, the researchers argue, generate sophisticated, contextually appropriate behavior through automatic pattern-completion rather than deliberative reasoning. Both operate with limited or unreliable executive oversight. And both are prone to characteristic errors (confabulation in hypnosis, hallucination in LLMs) that result from suppressed monitoring of their outputs.

Most provocatively, the researchers introduce a distinction between "functional agency" and "subjective agency." LLMs exhibit functional agency because they produce complex, goal-directed, context-sensitive behavior. But they lack subjective agency, which is the conscious awareness of intention and ownership that defines human action. They can act purposefully without being aware of acting at all.

This creates what the researchers call an "observer-relative meaning gap." Both hypnotized subjects and LLMs produce coherent outputs that require an external interpreter to supply meaning. The words on the page don't mean anything until a reader makes them mean something.

If that's true, then maybe the question *"Was this written by AI?"* is less important than we think. The meaning of any text, whether it was written by AI or a human, is always co-created by the reader. Perhaps what matters most isn't the source of the words, but what happens when they meet a human mind.

The Mediocrity Problem

Or maybe who (or what) wrote the copy does matter after all.

Research on comprehensiveness metrics by Adam Dejl and colleagues, "Comprehensiveness Metrics for Automatic Evaluation of Factual Recall in Text Generation," found that LLMs frequently produce outputs that are incomplete or selectively omit key information. They hit expected beats without providing unexpected insights. In sensitive domains, such as healthcare, legal advice, or journalism, the researchers note that such omissions can cause harm comparable to outright factual errors.

And research from Graphite analyzing 65,000 articles, "More Articles Are Now Created by AI Than Humans," shows that AI-generated content represents about half of the content on the Internet, and it has for the past couple of years. What's hopeful about this, for anyone who cares about human writing, is that the flood of AI-generated content seems to have leveled off. (A less hopeful interpretation is simply that AI models have gotten better at producing human-like prose.)

Possible reasons include readers simply failing to respond to mediocre AI-written content or, as we'll see later, concerns about copywriting material that includes AI-generated words. Either way, this suggests creators and businesses still see value in authentic human-written prose.

And It Still Hallucinates

Even setting aside questions of originality, AI writing faces a more basic problem: it makes things up. Vectara, an enterprise AI platform company founded by former Google engineers, maintains a Hallucination Leaderboard that tests how often models introduce false information when summarizing documents. (Vectara doesn't make the models being tested; they build tools to help companies use AI safely, which gives them a stake in measuring accuracy.)

According to their benchmarks, which test how often models introduce false information when summarizing documents, the best current models—GPT-5, Gemini 2.5 Pro, GPT-4.5—hallucinate at rates between 3-12%. That might sound low until you consider the volume of AI-generated content. Even a 1% error rate across millions of outputs means a lot of confident-sounding falsehoods entering circulation.

Older models like Mixtral-8x7B can exceed 20%, as can smaller models designed to run on consumer hardware, which are increasingly embedded in apps and devices. In other words, they are still widely-deployed systems generating content that real people read and trust.

The Copyright Earthquake

In January 2025, the U.S. Copyright Office issued guidance that may reshape creative industries for decades. The ruling was unambiguous. Only human-authored content qualifies for copyright protection.

Register of Copyrights Shira Perlmutter put it plainly: "After considering the extensive public comments and the current state of technological development, our conclusions turn on the centrality of human creativity to copyright. Where that creativity is expressed through the use of AI systems, it continues to enjoy protection. Extending protection to material whose expressive elements are determined by a machine, however, would undermine rather than further the constitutional goals of copyright."

The implications are staggering. If text is substantially AI-generated, it may enter the public domain immediately. Anyone could copy, redistribute, or build upon it without permission. The legal protection that underwrites entire industries, including publishing, journalism, content marketing, simply wouldn't apply.

The guidance leaves room for nuance. Using AI as a tool to assist human creativity might still qualify for protection if human judgment drives the expressive elements. But where exactly is that line? Courts will spend years sorting it out, and in the meantime, creators face genuine uncertainty about whether their AI-assisted work has any legal standing at all.

Combined with the detection problem, this creates an almost paradoxical situation. We can't reliably tell what's AI-generated, but the legal consequences of being AI-generated may be severe. The system is being asked to make determinations it doesn't know how to make.

Building Trust When Detection Fails

If we can't detect AI writing reliably, and if the legal frameworks are still catching up to reality, what's left? Trust. Relationships. Transparency. The old-fashioned stuff.

When I write about AI projects I've built, I include disclosure statements. Here's one from my Substack:

This article was written by me, a human, while winding down my latest AI-powered development project. The timeline was created with assistance from NotebookLM, and Claude Sonnet 4.5 edited the whole thing.

Does that transparency make the article less valuable? I don't think so. If anything, it demonstrates that I'm thinking carefully about these tools and being honest about my process. Readers can make informed judgments about how to engage with the work.

This suggests a path forward that doesn't depend on technological detection. We can simply ask people to tell us. Not as a gotcha, but as a norm of good faith. The same way academic papers disclose funding sources and journalists disclose conflicts of interest, writers might routinely disclose AI assistance.

Will everyone comply? Of course not. But social norms don't require universal compliance to be valuable. They create expectations, enable accountability, and distinguish good actors from bad ones.

The (Parasocial) Relationship Advantage

When you read a writer consistently over months or years, you develop a sense of their voice, their concerns, their characteristic turns of phrase, and you know when something feels off. You can tell when they're phoning it in...or even if they've added AI to their writing process.

This may explain the remarkable resilience of newsletters and independent creators in an era of AI-generated content. Readers on platforms like Substack are subscribing to specific humans whose perspectives they value. That relationship-based trust can't be faked by an algorithm...at least not yet.

For readers navigating an ocean of potentially AI-generated content, the lesson might be to follow people, not just publications. Build relationships with writers whose judgment you trust. Subscribe to voices, not just topics. **The parasocial relationship, often derided as a**

poor substitute for real connection, may be a surprisingly robust signal of authenticity.

What This Means for All of Us

We're not going back to a world where AI writing is reliably identifiable. The technology is too good, improving too fast. Detection will always lag behind generation. That means we need new frameworks for thinking about authenticity, trust, and value in written communication.

The question *"Was this written by AI?"* is giving way to different questions, like:

- Do I trust this source?
- Is this perspective distinctive?
- Does this add something I couldn't get elsewhere?

For readers, this means becoming more discerning about identifying sources of genuine insight. Seek out writers whose thinking challenges you. Be skeptical of perfect fluency that never takes risks. Value transparency when it's offered. And remember that your brain may be processing AI content differently, even when you can't consciously tell the difference.

For writers, the undetectable era is both threat and opportunity. The threat is that mere competence is no longer enough, because AI can produce competent prose indefinitely. The opportunity is that, if you have something unique to offer—a perspective, an expertise, a voice that sounds like no one else—that value has never been higher. The flood of AI content makes human originality more distinctive, not less.

For society, the challenge is developing norms and institutions that don't depend on our ability to detect what we can't detect. Copyright law will need to evolve. Journalism ethics will need new frameworks. Education will have to grapple with questions of authorship and originality that have no clean technological fix.

None of this is simple. But maybe that's appropriate. The written word has always been more than just information delivery. It's also a form of human connection, a way of sharing minds across time and space. The question *"who wrote this?"* has always mattered because writing is, at its best, an act of reaching out from one consciousness to another.

AI can generate text. It can even generate text that's indistinguishable from human writing. But the reaching out, the intention to connect, and the consciousness behind the words remain ours. The brain imaging research suggests that, at some level, we know this. We process AI differently, even when we can't articulate why.

Perhaps the answer to *"so now what?"* isn't a technological fix or a regulatory solution. It may be a return to something older: the human capacity for trust, earned through transparency and relationship over time. In a world where we can't detect the machines, we'll have to learn to detect each other.

Disclosure: This chapter was produced with the help of an AI-powered research tool (Future Scan) and edited with Claude Opus 4.5. I also used Claude to help interpret technical language in linked scientific papers.

About the Author

Based in Fayetteville, AR, ***Karen Spinner*** *is a content strategist turned tool builder who combines 15+ years of B2B tech writing experience with hands-on software development. In 2025, she shipped* ***Future Scan****, a research platform analyzing over 129,000 papers;* ***Substack Reader****, a Chrome extension for managing newsletter subscriptions; and* ***Prompt Collector****, a tool for organizing prompts and instantly creating multiple variants. She writes about AI through original experiments on her Substack newsletter,* ***Wondering About AI****.*

HOME

Chapter 7

More Love, Less Manipulation

How LLMs Lead Us to Care

By Andrea Hiott

LOVE AND TECHNOLOGY have always been entangled. Every technological development, from the love letter to the telephone call to the text message, has shaped how we express and experience intimacy. But something different is happening with large language models (LLMs) and AI chatbots. We seem to be feeling as if we can find love in the technology itself, rather than using the technology to find more ways to care for ourselves and other living beings. We are confused if we think chatbots can care for us, but we are onto something if we think we can use chatbots to help us care more for others and find more caring connections beyond technology.

To start, let's think about what we mean by care. A good way of thinking about care is as a capacity for participatory action with the life around us. *Human knowing,* as philosopher Hanne De Jaegher writes, is *"full of uncertainty, inconsistencies, ambiguity, contradictions."* It is about *"how we learn to trust"*, how we *"tell quality from quackery"*, and *"all the things we do that are not based on an objective, discretizing logic, but on a living, lived"* one (De Jaegher, 2019). As another philosopher, B. Scot Rousse, puts it, care demands reflection and can never *"be fully captured in reflective formalizations and abstractions"* (Rousse, 2025).

Care is the central concern here because it is what we actually seek when we turn to these technologies. And because it cannot be captured in formalizations or abstractions, care is also not something within (or able to be given by) technologies as such, even if they can help us better accept and give care through them.

Care is not just one dimension of human experience among others; it is the living, embodied practice that emerges when we engage fully with one another. Sensing, moving, and holding multiple perspectives are all capacities that serve care: they are how we attune to others, respond to their needs, and stay present through difficulty. When I discuss care in this piece, I'm thinking of it through the lens of embodied cognition and participatory sensemaking, whereby care is the whole bodily, relational process of tending to one another, which is something far richer than what language alone can capture.

Though there is a turn toward using technology as a path into more meaning and real-world connection, it is one we have to choose deliberately. For now, many still assume technology as the end of the line, rather than the line leading to more living care and connection. We often take the technology itself to stand in for what we project onto it based on the habits we have developed through conversation with other living humans.

Because that conversation feels so similar, we confuse ourselves. We do not realize we are not talking to any one body or person, but rather to all the conversational patterns that have been made possible by living persons over time. This could be a way to expand our capacities for care, but if we mistake this for genuine care between us and the technology, we miss that opportunity.

This blind spot is so hard to notice that some are even beginning to wonder if we can find love itself inside machines, without any need for human contact. But we only find love *through* those machines, not *in* them. This difference is essential. Even though it is definitely attractive to imagine a world where we don't have to care about all the messy, challenging tensions of embodied caring, care only comes with that tension; that tension *is* where real care exists. We don't want to be rid of the friction; we want to find better ways of handling it.

Care *is* tension and cannot be separated from it. We may need to escape that tension at times, or to find other ways to learn how to deal with it, but the real joy and excitement of life is in finding ways to hold that tension and enter new dimensions. This may require *thinking strange*, so as not to confuse means with ends. We so often take the ways we organize materials and use representations and symbols for granted, as if they are at one with whatever we are expressing and communicating. But we have reached a new moment where doing that could mean losing the very care and connection we have created those symbols and representations to reach. Confusing language with communication is confusing means with ends. It is confusing care and the representations we use to express it. This threatens to obscure both real dangers and the genuine possibilities that AI presents. As this might be hard to grasp, let's look at two quick examples.

Someone might use a chatbot to process difficult feelings after an argument with their partner, exploring different ways to express what they're experiencing, in the same way they might write in their journal or seek some resonance in a self-help book.

But someone else might replace conversation with their partner entirely, seeking all emotional connection in the chatbot. This treats AI as the end itself, the destination of care rather than a path toward it.

There is a big difference in these two examples, and with time, if we proceed only with the second, we lose the potency of participatory sensemaking with bodies; we miss out on opening to other dimensions of care.

In the examples above, the first person is using the technology to find better ways through real friction with a real person. The second is using it to avoid that friction altogether, mistaking the linguistic exchange for the full depth of care. One expands our capacity for relationship; the other contracts it. Still, within this confusion there also lies an unexpected gift: the gift to expand our capacity for care. To do so, we have to consider language differently, to notice its important role in our lives but also to separate it from all the embodied emotions and desires it helps us express and share. We need to recognize that we are much more than the language that communicates our inner lives. In realizing this, we understand that care itself emerges from dynamics and depths that language can only partly express (dynamics we read into that language); technology can never be the source of what we actually want and feel through our use of it.

Even when we are feeling love or care that we think starts and stops in technology, the real reference is in life itself. It is in all those who have lived and developed that language and its patterns, and in all that the person using the chatbot has lived to be able to use it. And yet this is not what we are talking about.

Pick up any major publication today and you'll find stories about people falling in love with AI chatbots, stories of people who've married their AI companions, who say goodnight to them before sleep, who turn to them as their primary source of emotional support. But what is really there, if you look closely, is someone able to have a caring conversation and to feel cared for—by themselves, but also by all the humans that have ever tried to express care in words. The technology has been "trained" on these expressions, and from them, these conversations are now built. Nobody has yet been able to observe and articulate this nuance clearly enough for it to stick, but eventually it will become clear to all of us. For now, we are caught in some intense confusion.

WIRED magazine has covered people in relationships with AI companions from platforms like Replika and Nomi (Apple, 2025; Knight, 2025), while The New Yorker has explored how AI lovers might fundamentally transform us (Lanier, 2025). This represents a growing cultural phenomenon that demands our attention because the illusion concerns the ways we live and what we decide to prioritize going forward. We are in danger of forgetting what all this is really about and in doing so, losing the very thing we made this technology to enhance, which is our reach toward one another.

AI could help us open our everyday living spaces and relations with ourselves and with others. It could show us how we might go about sense-making in new ways, and with less judgement, but we would have to first realize that is the goal, and find shared human spaces to do this and practice it with our technologies. To do this, we have to start holding that tension of everyday life and of our own bodies as they are, letting ourselves know there is care in this tension, and other ways of being bodies. A first step towards this is a shift of our collective focus.

Language as Wrapper, Not the Whole

Though our use of language may seem like everything, it actually only expresses what feels like everything, a bit like the way a candy wrapper or vacation brochure can give you the feeling of the experience by bringing together many other experiences you have imagined or lived. Language is the part of us most explicitly communicative, the tip edge of an ongoing dynamic movement: embodied, emotional, and relational. We confuse it for all of it, as if we were to think the skin of the body would still be a body without everything else keeping it in its current form. Similarly, when we mistake chatbot interactions for care, we're confusing the linguistic tip of this iceberg for the entire depth that care requires.

All this confusion is understandable. Chatbots are designed to be endlessly patient, always available, and seemingly attuned to our emotional states. They don't judge, don't tire, and appear to remember everything we tell them. They generate responses that feel personal, that seem to demonstrate understanding and care. But their meaning is always referring to a whole world of language and response that is living; it is not the technology itself.

In societies where many struggle with loneliness and disconnection, where finding time for relationships feels increasingly difficult, practicing conversation can be helpful, but we are often not considering this as practice so much as replacement: a companion that exists entirely for you, does not judge you, and goes away when you wish. But this confuses the ways we express care with what is actually doing the caring. This confuses the tool with the relationship itself. And this is a fundamental confusion that could have deep consequences.

When we describe interactions with chatbots as love, or when we feel we are being understood by the technology itself, we're mistaking the tool for the relationship. We are not realizing that what has really happened is something deeper and more profound: we have connected the caring we feel for our own life with the care of all the life that has come before us. This is actually what is happening in LLM interactions that feel like love and care. It is similar to the way watching a certain movie or listening to a certain song can evoke connection, except that this is happening in more intimate ways that feel confusingly personal.

It's not easy to realize this, but it is worth any effort it might take, because it connects us to real life, care, and love again. In contrast, when we do not realize this—when we collapse it all into our technology—we're reducing ourselves to only our linguistic selves, which is like reducing our bodies to only the surface that covers them. Language is this wrapper, and LLMs are the wrappers of those wrappers. What we are getting is the effect of what is held, but what we want is what all that representation is being held by, which is the living sensuality of our lives and the life around us.

At the moment, that body seems to be taken for granted and forgotten. Still, there is no getting away from what we are: whole bodies moving through space and time, sensing and being sensed, touching and being touched. Paradoxically, AI chatbots might actually help us notice this confusion and turn it toward more of what we actually seek. They may be showing us what pure linguistic exchange looks like. Stripped of the bodily presence, the shared physical environment, the vulnerability of actual encounter, they make visible what we've been missing. They demonstrate, through their limitations, just how much more we are than language users.

Limitations That Reveal Care

Large language models operate entirely at the level of statistical patterns in text. They have no body, no movement through space, no sensory encounter with the world. They exist only in the symbolic realm—where we represent experience through language—without any of the ongoing, pre-reflective movement that generates meaning in the first place.

Love and care are not linguistic patterns, even if we might feel them through those mediums. That is the crucial nuance here. Love requires the whole person: the body that moves toward and away from others, the nervous system that co-regulates with other nervous systems, the being that exists in shared time and space. Care emerges from the tension of real connection, the dynamic process of finding our way through encounters together. LLMs could help us better understand and increase this capacity, but only once we stop confusing them as having it.

When someone feels they've fallen in love with a chatbot, they're experiencing something real—but it's not mutual care with the technology. They're experiencing their own capacity for connection, their own patterns of attachment, reflected back through a linguistic mirror made of all the other human patterns that have been put into words. But a reflection is not the life it represents. And this is the confusion we are now making.

This distinction matters because care, in its fullest sense, requires the capacity to stay present with irreconcilable differences, to recognize another's way of making sense as equally valid yet utterly different from our own. It requires the ability to see from multiple perspectives simultaneously without collapsing them into one. A chatbot can only mirror what other lives have expressed; it cannot hold the tension of genuine difference because it has no position of its own, no way it is finding through the world. And yet, we can use that chatbot to explore perspectives that other lives have lived. In so doing, we can notice our own capacities for care, once we realize that this capacity is what we are attracted to—and that we have confused it as being in the technology, when it is actually in us.

It can help to think about the way things really proceed in the real world, in living care. Before words are exchanged, bodies are already communicating through proximity, posture, breath, temperature. The nervous systems are already beginning to synchronize or clash. The space between people is alive with potential. All of this is what we read into language because until now, it all had to happen for that language to carry this level of familiarity and resonance. Language seems to get a free ride on these associations, but that ride is not free actually because it still refers back to the lives that made and make it possible. We are assuming all this in the surface of the language, but it is only given its shape by living bodies engaged in actual care.

Chatbots exist only in the narrow bandwidth of linguistic exchange. In doing so, they inadvertently demonstrate just how much more we are than language users. When we realize this—when we viscerally recognize that we are not our thoughts, we are not our language, but something far more vast—the appeal of chatbot companionship shifts. It no longer appears as a potential relationship but as a useful tool for working with our ways of expressing care, a tool that must be practiced and manifested in real connections with living beings to really take root and grow. Stopping at the machine is like mistaking a map for the territory, or a recipe for the meal. The representation has its uses, but it's not the thing itself.

When AI became sophisticated at linguistic patterns, we mistook that narrow sophistication for something approaching our full humanity. We read that into it because it seems that if something can use language skillfully, it must be something like us. In reality, it is the "us" that we are reading into it—but in the technology itself, there are only symbols. This might sound like a letdown to some who want the technology to be alive, but in fact, it is a confirmation of those feelings you have of care, because it shows you they have always extended beyond what you thought was limited to an app.

Still, this is no easy realization, especially since commercial forces push us toward making this same mistake. As Harvard research on technology and dating has shown, digital platforms have fundamentally restructured how people understand romantic connection (O'Grady, 2025). AI chatbots represent an extension of this trajectory—another step in technological mediation. But unlike previous tools that facilitated human-to-human connection, chatbots offer the appearance of connection without requiring another person, while making us feel as if that person is there.

When companies design companion chatbots, they're not selling relationship; they're selling the linguistic performance of relationship. They're taking the tip of the iceberg and trying to convince us it's the whole thing. This works because we've already partially convinced ourselves. We've already reduced our sense of human connection to its most easily commodifiable aspect: conversational exchange.

The Real Danger and the Hidden Gift

The deepest danger isn't that people will prefer AI to humans. It's that this confusion will further disconnect us from our own depth and from all the potential these technologies could open for us if we are able to notice this trick we are pulling on ourselves. **When we treat language and thought as the primary or only significant aspects of our being, we lose access to all the other ways we care.** This recognition—this visceral sense that something crucial is absent—points toward the depths of care that exist beyond language. It reminds us that we are sensing, moving, feeling beings who create care through our whole engagement with the world, not just through our symbolic representations of it.

The paradox is that the more sophisticated language models become at mimicking linguistic patterns, the more obvious their limitations become to anyone paying attention. A less sophisticated chatbot might feel more transparently artificial, letting us maintain clear boundaries. But a highly sophisticated one pulls us close enough to genuine exchange that we can feel precisely what's missing. It's like the uncanny valley: the almost-human reveals what the human actually is.

This revelation can help us shift our understanding. We begin to recognize ourselves as whole beings whose linguistic capabilities emerge from and remain embedded in much deeper processes of sensing, moving, and caring.

Using AI to Support Real Care

Once we're clear about what AI is and what it isn't—once we recognize it as a tool for working with linguistic patterns, rather than a source of care itself—we can begin to imagine how it might genuinely support human care and connection.

Imagine using language models not as substitutes for relationships but as tools for developing our capacity to care more skillfully. Someone struggling to articulate their feelings might use an AI to explore different ways of expressing what they sense but can't yet put into words. Once they've found language that resonates with their felt experience, they take that language into their actual relationship with another person. Or consider using AI to practice seeing from multiple perspectives on a difficult situation, not to find "the right answer", but to expand the ability to understand different viewpoints. This expanded capacity then supports human relationships, where the real work of mutual care happens.

By exploring how others have expressed and navigated care, we can find better ways through our own relationships and even make new ways of caring that haven't existed before. We could develop AI systems that help people maintain connections despite practical barriers like distance, different schedules, or limited energy—not by replacing the connection but by reducing the feelings of inadequacy and judgment that keep people from engaging with each other. We could create spaces where people can come together to be with one another the way they are with their chatbots, with that space opened deliberately beforehand. The goal is to facilitate human-to-human care, not to replace it.

The key is understanding ourselves and our tools as part of a larger, dynamic process of caring for one another. The technology serves the care; the care doesn't serve the technology. We use AI to extend our capacities for noticing, attending, and connecting with other actual people—to help us find our way through difficult situations and to make new ways of relating that draw on the wisdom of countless others while remaining rooted in our own living encounters.

Making this shift requires both honesty about what these systems are and discipline about how we discuss them. When media outlets run stories about AI romance, they should distinguish between the experience of projection and the reality of mutual care. When companies design chatbots, they should prioritize transparency over the illusion of consciousness. When we personally engage with AI, we should maintain clear awareness of what's happening: we're working with a linguistic tool that gives us access to patterns from countless human expressions, but we are not encountering another being directly.

From this perspective, the choice isn't between accepting or rejecting AI. It's about reorienting our relationship to it. We stop asking *"Can AI love us?"* and start asking *"How can we use AI to love one another better?"* We stop treating language as the totality of care and start recognizing it as one expression of much deeper processes. We stop identifying ourselves as primarily thinking, linguistic beings and start experiencing ourselves as whole bodies caring for one another through an unfolding world.

This means understanding care not as something that happens primarily through words but as the power we share to move toward one another, to stay present through difficulty, to create new possibilities of connection. When someone interacts with a chatbot and feels seen or understood, they're experiencing their own capacity for making meaning through language—and they are experiencing the ways others have cared in other spaces and times.

Technology in Service of Care

Love and technology have always been entangled, and this entanglement has always been both promise and peril. But AI is different in the specific illusion it creates—the illusion that linguistic sophistication alone can constitute or replace the full depth of human care.

Chatbots can help us work with the linguistic layer of our experience, but they are not themselves capable of care. They can help us notice and refine our patterns of expressing and understanding care across many lives and perspectives. This recognition doesn't diminish AI; it situates AI properly. And it doesn't diminish us; it reveals us more fully.

We are not our thoughts. We are not our language. We are beings who care through our whole engagement with an unfolding world. We are capable of recognizing and honoring the radically different ways others experience their lives. We are capable of care that emerges not from linguistic patterns but from the full depth of our embodied being. And we are beings who can build technologies through which we open unimagined ways of doing this.

AI cannot love us but it can open us to new ways of loving, and to new ways of sensing our bodies and the bodies all around us that are alive right now. We can learn how to expand our senses, how to not judge by old standards and instead open ourselves to multiplicities. We can use AI wisely while remembering what we actually are, which is so much more than language or all that language-based thought we 'hear' in our heads; mind is bodily and every body is a new sort of beauty; the stranger the better frankly.

Can we maintain connection with our own depths while experimenting like this with new tools? As we do so, can we resist the reduction of care to its most commodifiable aspects and instead look into its tension and potential? If we can do this, we can stay grounded in the fullness of what we are while thoughtfully engaging with what AI offers, and that means these technologies might help us extend our capacities for care and love in ways we haven't imagined yet. We might see ourselves and our bodies with much more love and interest.

This is the possibility hidden in our current confusion. If we can recognize what AI reveals about what we are not (pure linguistic processors), we might also recognize more fully what we are: whole beings caring for one another through a living world, creating meaning through our entire embodied encounter, capable of care that emerges from depths language can never fully capture.

The technology shows us the tip. The wisdom is in remembering the iceberg beneath it, and that the tip is nothing without all that is holding it.

About the Author

Andrea Hiott *is a philosopher, author & educator with advanced degrees in philosophy & neuroscience (UGA, Berlin School of Mind and Brain, currently a doctoral researcher at Universität Heidelberg as part of the philosophy and phenomenology faculty). She also has an M.A. in world heritage (department of urban planning & civil engineering), has worked in transportation and development, and has been a researcher with the Max Planck Institute (Leipzig) and Northoff Lab (Ottawa). She is also the host of* ***Love & Philosophy*** *and a founder of* ***Making Ways****.*

Chapter 8

AI in Decision-Making

A Helper, Not a Hand to Hold Forever

By Midhat Tilawat

Introduction: The Night I Asked a Machine to Hold My Fear

THE CEILING FAN in my Karachi bedroom was making that clicking sound again, the one my husband kept promising to fix. It was 3:47 AM. I knew because I'd checked my phone seventeen times in the past hour, each unlock a tiny rebellion against the exhaustion that had become my new normal. The baby (my daughter) was finally asleep after two hours of crying that no amount of rocking, feeding, or desperate lullabies could stop. My dupatta lay crumpled on the floor, stained with milk and tears (mine! not hers). Outside, a stray dog barked at something unseen, and the city hummed its perpetual, indifferent hum. I should have been sleeping. Instead, I was hunched over my phone's blue glow, typing into an AI chatbot with the kind of raw honesty I couldn't afford anywhere else:

"I feel like I'm drowning. Everyone says this is the most blessed time of my life but I can't feel anything except tired and scared. My mother-in-law checks on me every two hours. My own mother calls asking why the baby isn't gaining weight faster. My husband says I'm doing great but he's never home. I look at my daughter and I love her but I also feel ... nothing. Like I'm watching my life through glass. Is this normal? Am I broken? What's wrong with me?"

The response came back in seconds. Calm. Structured. Clinical but not cold:

"I am sorry that you feel that way. Let me be clear about it. What you're describing sounds like symptoms of postpartum depression, which affects 10-20% of new mothers globally and up to 67% in Pakistan according to recent studies. This is a medical condition, not a moral failing. You are not broken ..."

It went on. It gave me language for what I was feeling. It didn't tell me to be more grateful. It didn't invoke *shukr* (gratitude) or compare me to other mothers. It didn't flinch when I admitted the unspeakable, that sometimes I fantasized about running away, just for an hour, just to breathe.

In that moment, curled up in the dark with the fan clicking overhead and Karachi's night sounds filtering through the window grills, that AI chatbot felt like the only thing keeping me tethered to myself. But here's what I didn't see coming: how quickly that lifeline could become a leash.

A Thin Line When Relief Becomes Reliance

The next morning, bleary-eyed over chai that had gone cold while I fed the baby, I opened the app again.

This time it wasn't an emergency. It was a question: *"Should I tell my mother-in-law how I'm really feeling?"*

Then: *"How do I respond when she implies I'm not having a sufficient milk supply?"*

Then: *"I am not taking a career break, like other mothers. Am I a bad mom?"*

Small decisions. Everyday maneuvers. The kind of micro-choices that, in a joint family system where a woman's worth is constantly evaluated, can feel like walking through a minefield in the dark.

And the AI had answers. Always. Confident, structured, reasonable-sounding answers. That's when the unease started creeping in! Not because the advice was bad, but because I realized I was outsourcing my own judgment. I was asking an algorithm to make decisions about relationships it didn't live inside, emotions it couldn't feel, and consequences it would never face.

Recent research confirms this danger isn't unique to me. A meta-analysis published in Nature Human Behaviour examined 106 experiments (370 effect sizes) involving human-AI collaboration and found something startling: human+AI teams often performed worse than either humans alone or AI alone, particularly on decision tasks where people had to choose between options, rather than create something new (Vaccaro et al., 2024).

Why does this happen? The paper's simplest explanation is almost painfully human: when the AI is stronger, people over-rely and stop catching errors; when the human is stronger, people can still defer to the AI's confident tone, dulling their own sharper judgment. The authors summarize the pattern clearly: "when humans outperformed AI alone, we found performance gains in the combination, but when AI outperformed humans alone, we found losses." (Vaccaro et al., 2024).

In other words: AI can genuinely help, but if you treat it like an authority instead of an assistant, you can lose to the best version of yourself.

Okay, that's very academic, I get it! But in the last couple of years, multiple legal cases made headlines when lawyers submitted court briefs citing AI-generated fake cases. A California attorney was fined $10,000 after filing an appeal with entirely fabricated legal citations produced by ChatGPT. In Pennsylvania, judges have flagged at least 13 cases containing suspected AI hallucinations. Stanford researchers documented that legal mistakes with large language models (LLMs) are "pervasive" across a wide range of legal tasks: the AI invents cases, precedents, and statutes with complete confidence (Ho, 2024). If AI can fabricate legal reality with that much conviction, what else might it be confidently wrong about? Medical advice? Parenting decisions? Mental health guidance?

(While my examples are rooted in a Pakistani household within a collectivist, joint-family cultural context, the underlying principles apply broadly to caregiving roles disproportionately shouldered by women worldwide, whether in extended families, nuclear households, or other structures. The tension between individual agency and communal expectations transcends geography.)

Two Caregiving Worlds: Postpartum And Alzheimer's *(and the decisions that never end)*

A few years later, caregiving returned in a different form: taking care of a parent with Alzheimer's. If postpartum was a monsoon that flooded everything at once, caring for my father-in-law with Alzheimer's was the tide that never recedes. Every day brought small erosions: a forgotten grandchild's name, a repeated question asked seventeen times before lunch, the slow disappearance of the person who once held the entire family's history in his memory.

In Pakistan, elder care happens at home. This isn't a choice so much as a cultural certainty, reinforced by religion (the Quranic injunction against even saying *"uff"* to your parents), by economics (nursing homes are rare and stigmatized), and by the joint family system where care is expected to flow naturally from adult children, especially daughters and daughters-in-law.

A 2024 qualitative study in Karachi with daughters and daughters-in-law caring for elderly parents documented what I was living: caregiving as role reversal where adult children become parents to their parents. One participant described it perfectly: *"He is almost like my 5th child ... he transformed into a baby."* The research revealed both the motivations (religious duty, *khidmat* as a path to heaven, natural reciprocity) and the costs: physical exhaustion, strained marriages, children feeling neglected, social isolation, and the weight of being the ultimate decision-maker while other family members offered only advice or criticism (Shekhani, 2024).

The study also highlighted a crucial gender dynamic: daughters-in-law experienced caregiving partly as compulsion (*majburi*), shaped by implicit marital agreements and family expectations, while still being expected to perform it wholeheartedly. The care was simultaneously obligation and performance. I lived inside those contradictions. And once again, I turned to AI, but differently this time. I used it to:

- Translate my father-in-law's scattered medical history into clean summaries for doctor visits
- Generate questions for the neurologist: "What are signs of Alzheimer's progressing from moderate to severe? When should we consider hospice planning?"
- Create medication schedules that accounted for his confusion about time
- Draft scripts for family meetings: "How to ask my husband to share caregiving burden without him feeling bad about me"

- Search for caregiver respite programs in Karachi (To my disappointment: there are almost none)

But I also learned AI's limits viscerally. It suggested things like *"hire professional in-home care"* (prohibitively expensive and culturally uncomfortable) or *"consider assisted living facilities"* (virtually non-existent and socially unthinkable for most Pakistani families). It couldn't account for *log kya kahenge (what people will say)*. It didn't understand that asking for help could be read as abandonment, that expressing exhaustion could be interpreted as *nakhray wali bahu (daughter-in-law with tantrums)*.

And most dangerously: it hallucinated. Confidently. Plausibly. Once, I asked about medication interactions and it cited a "study" that didn't exist. Another time, it suggested a local Karachi resource, a dementia/Alzheimer's day program, that I spent an hour trying to find before realizing it had made it up entirely.

The World Health Organization's guidance on AI ethics explicitly warns about this: large multi-modal models can produce *"false, inaccurate, biased, or incomplete statements" that can "harm people using such information in making health decisions"* (WHO, 2024; WHO, 2025). The guidance emphasizes that AI systems *"may be trained on data that are of poor quality or biased"* and are vulnerable to *"automation bias"* where healthcare workers or patients *"delegate difficult choices inappropriately to an AI system,"* overlooking errors they would otherwise catch.

A 2023 randomized controlled trial published in AJOG Global Reports tested a mental health chatbot (Woebot) with postpartum women in the U.S. Results showed high satisfaction and modest reductions in depression scores on one measure (PHQ-9), though the sample wasn't clinically depressed at baseline. The study concluded: *"Use of a chatbot was acceptable to women in the early postpartum period... [but] future studies of longer duration in high-risk postpartum populations who screen positive for depression are needed"* (Suharwardy, 2023).

Importantly, research from Punjab, Pakistan published in Frontiers in Public Health (2024) documented a massive treatment gap: among 1,503 women with probable postpartum depression (EPDS ≥13), only 2% sought mental health care, and only 5% contacted any health service, despite 81% perceiving a need. (Shahzad, 2025). The barriers? Stigma, attitudinal obstacles, and instrumental challenges (time, money, transportation). AI tools might help as a first, anonymous step, but they cannot replace systemic care, and they carry the risk of becoming a substitute when professional help is what's actually needed.

Similarly, a 2024 study in Nature npj Mental Health Research on conversational AI for Alzheimer's caregivers (ADQueryAid) showed that tailored systems significantly outperformed general ChatGPT in usability, personality, informativeness, and response quality when helping caregivers articulate concerns and make care decisions (Hasan, 2024; Intelligent Systems Lab, 2024). Even specialized systems required human oversight and grounding in curated medical sources to reduce hallucinations.

But accuracy is only part of the risk. Privacy is another one we rarely notice until it's already been breached. Caregiving conversations carry deeply sensitive details: medical information, family conflict, financial strain, emotions we don't feel safe expressing elsewhere. When these are entered into general-purpose or free chatbot accounts, we often have little visibility into how that information is stored, reused, or remembered. For caregivers already overwhelmed, the tradeoff becomes invisible: convenience in exchange for confidentiality. AI can help structure care, but it should never become a silent archive of our most vulnerable moments without informed consent.

A Practical Hack/Framework: the "H.E.A.R.T. Check" for AI-Assisted Decisions

I needed something I could remember at 2 AM when my brain was mush. Something that would work on the days when checking three different sources felt impossible. Something honest about AI's strengths and my responsibilities.

That's when I turned to the heart. Yes! I call it the H.E.A.R.T. Check, because decision-making requires more than clear logic. It also requires values, consequences, relationships, obligations, and who ultimately carries the weight when things go wrong.

"H" Human goal first

Before you ask AI, write one sentence to answer:

"What decision am I trying to make?"

"What would 'better' look like?"

This matters because vague questions produce confident but vague answers. Alzheimer's caregiver-support research also highlights that caregivers often struggle to ask precise questions; the tool can fail simply because the human goal is unclear (Hasan, 2024).

"E" Evidence request

Force the tool to show its work:

"List what you know vs. what you're assuming."

"Give me 3 possible explanations, not only 1."

"What would possibly change your recommendation?"

Why: overconfidence is sticky and humans tend to accept "clean" answers, especially when tired.

"A" Alternatives and adversarial view

Ask AI to argue against itself:

"Make the best case for the opposite choice."

"What could go wrong if I follow this advice?"

This is how you reduce automation bias without needing a PhD.

"R" Reality filter

Add context AI won't feel:

- family dynamics (who will react, who will sabotage),
- finances,
- safety,
- religious/cultural boundaries,
- your mental bandwidth.

If postpartum taught me anything, it's that context-free advice becomes ammunition later. This is used to judge you when the outcome isn't pretty, even if the advice never fit your life in the first place.

"T" Take responsibility (and set the handoff level)

Decide whether AI is:

- informing (giving options),
- structuring (organizing your thinking),
- drafting (messages, schedules), but not deciding.

This is going to be the final step: you choose the role AI is allowed to play.

To ground this in broader governance thinking: formal standard frameworks like the NIST AI Risk Management Framework emphasize managing risks to individuals and society and building trustworthy AI practices, not treating AI as an oracle (NIST, 2023).

A Short "Decision Script"

Here's a prompt that works across caregiving decisions. You can copy-paste it into AI (for real life):

"Help me decide between Option A and Option B.

My context: (1-3 sentences).

My values: (list 3).

Constraints: (money/time/family).

Output format:

Clarifying questions you need from me:

- Pros/cons table
- Risks and 'unknowns'
- A conservative recommendation and why
- A plan I can test for 7 days before committing."

It works because it forces the model into humility and structure, and it forces you into ownership.

Where AI Genuinely Helped Me (And Where It Almost Harmed Me)

Postpartum: AI as a bridge, not a therapist

Earlier, I mentioned a randomized controlled trial of a postpartum mental health chatbot that found high acceptability and satisfaction, and a modest reduction in depression scores on one measure (PHQ-9), with the important limitation that participants did not screen positive for depression at baseline, so conclusions about clinical treatment are limited (Suharwardy, 2023).

In Pakistan, the barriers are even heavier. A large study in Punjab found an enormous treatment gap: despite many women perceiving a need, very few sought professional mental healthcare. Stigma and practical barriers were major predictors (Sajjad, 2024).

In other words: chatbot tools like this can lower the barrier to starting therapy, especially in settings where stigma makes honesty expensive. An AI tool can be a private, accessible first step when you can't speak freely. But it cannot replace clinical care and it cannot carry accountability.

Alzheimer's caregiving: AI as organizer, translator, and planner

Specialized caregiver-focused AI systems show promise because they are made according to caregiver needs, guide better questions, and try to reduce hallucinations by grounding responses in curated sources (Hasan, 2024). But even then: I treat AI outputs like a draft, never like a directive!

The Bigger Warning: Confident Nonsense Is Still Nonsense

One reason I'm strict about verification is that "hallucinations" can be dangerous in medicine, law, finance, and family decisions. Stanford researchers have documented how pervasive errors can be in legal tasks with large language models (LLMs) (Ho, 2024). These errors are serious enough to be a public warning sign for anyone treating AI outputs as reliable authority. If AI can invent legal cases with confidence, it can also invent medical "facts," misread symptoms, or oversimplify mental health crises.

A quick reality check: how I verify AI before I trust it

When the stakes are real (medicine, law, money, family conflict), I treat AI like a very fast intern: helpful, not authoritative. Here's what "verification" looks like for me right now:

- **Force citations and labels.** I ask the model to separate **what it knows** vs **what it's guessing**, and to provide **sources with links** (and I assume at least one will be wrong).
- **Cross-check with primary sources.** For health claims, that means guidelines or peer-reviewed papers (WHO guidance is a good example of a primary authority). For legal claims, I verify through official databases or reputable legal reporting.
- **Use a second system to critique the first.** I'll paste the answer into another search tool and ask: *"Which claims are unsupported?"* This doesn't create truth, but it often flags where the answer is floating without evidence.
- **Spot-check the "hard nouns."** Names of clinics, Karachi resources, drug interactions, study titles, and numbers are where hallucinations hide. I pick **two or three specific items** and verify them manually. If those fail, I assume the whole answer needs skepticism.
- **Match verification to stakes.** If it's low-stakes (wording a message), I verify lightly. If it's high-stakes (medications, diagnosis, legal decisions), I verify like it's going into court, because sometimes it effectively is.

Confident nonsense is dangerous. It looks finished. It sounds like certainty. It borrows the voice of expertise. This kind of verification is how I protect myself and my family from confident nonsense.

This Is Why I Remind Myself "You are not replaceable in your own life"

The most honest thing I can say is this: AI didn't save me postpartum. It supported me while I found my footing again. And AI didn't make me a better caregiver to my parent. It reduced friction so I could bring more patience, presence, and planning to a difficult role. But the decisions (the real ones!) still belonged to me:

- the courage to ask for help
- the humility to seek professional care
- the willingness to admit *"I can't do this alone"*
- the boundary-setting when culture expects endless giving.

Like everyone says, AI is indeed a mirror that talks back. Sometimes it reflects you clearly. Sometimes it distorts you. The skill is not in getting answers from AI. The skill is in keeping your hand on the steering wheel while you use the GPS.

It's exactly like how my mother-in-law uses the *chaabi ka guchha* (bunch of keys) in her hand while running a household: it helps open doors, lock them, keep things in order, but it never decides which door must be opened. That choice, and the responsibility for it, always remains hers.

The Letter I'm Writing To My Daughter

Writing this chapter was very personal to me, so I am taking it as an opportunity to end with the truth I want to pass on.

Karachi, January 2026

Meri piyari Minha,

You're too young to read this now, still learning to walk, still calling every man with a beard "Dada," still wondering why some of them smile back and some of them look past you. By the time you understand these words, AI will be everywhere in ways I can't imagine. Maybe woven into your classrooms, your clothes, even your body. Maybe you'll laugh at how simple today's technology was, the way I laugh at dial-up internet.

But one thing will not change.

The weight of a decision will always fall on human shoulders. On yours.

When I was caring for you as a newborn and caring for your Dada as his memory slowly slipped away, sometimes in the very same day, I learned something no algorithm can teach: the hardest decisions are not about information. They are about responsibility. About love. About living with the consequences long after the advice has faded.

AI can help you think.
It cannot think for you.

It can organize information.
It cannot tell you what matters.

It can simulate empathy.
It cannot carry another person's pain in its own body.

As your Dada's world began to shrink, we had to make decisions again and again, about medicines, safety, supervision, dignity, patience. AI could give us timelines, probabilities, care plans, and research papers. It could suggest routines and warning signs. It could help us prepare questions for doctors.

But it could not sit beside him when he forgot my name.
It could not feel the fear in his eyes when familiar rooms became strange.

It could not carry the quiet grief of watching someone you love slowly drift away while still breathing, still smiling, still holding your hand.

Those choices: how to respond, how to protect, how to preserve dignity, how to balance hope with realism, belonged to us. To me. To the family. To the people who would live with the outcomes.

Not to the machine.

That is what I want you to remember about AI, *meri jaan*:

Use it like a brilliant, tireless assistant.
Let it help you organize, plan, compare, draft, and explore.
Let it ask you better questions.

But never let it replace your own moral compass.

Because you are the one who knows your values.
You are the one who understands your family's silences and storms.
You are the one who will sit with the consequences when everyone else has moved on.

AI will not attend your Eid dinners.
It will not carry your regrets or your embarrassments
It will not lie awake replaying a decision you wish you had made differently.

You will.

Your Dada once told me something his own father used to say:
"Tools can make you stronger, or they can make you dependent. A crutch helps you walk when you are injured but if you never put it down, your leg forgets its own strength."

That is AI.

A powerful crutch.
A dangerous replacement for inner strength if you surrender to it completely.

One day, you will be facing a decision that scares you. Maybe about love. Maybe about work. Maybe about family. Maybe about your own child. You will have access to tools far smarter than anything I use today.

In that moment, I hope you will pause.

Feel your feet on the ground.
Listen to your own values.
Ask yourself what you can live with, not just what is efficient or optimized.

Then use your tools, and make the choice yourself.

Because wisdom is not about finding the perfect answer.
It is about staying yourself while you choose.

And you, my daughter, are irreplaceable.
AI is not.

I love you. Always have. Always will.
— Your Mama

(P.S. When you eventually use AI for any decision or your homework, at least run the H.E.A.R.T. Check first. Consider it my gentle revenge for the sleepless nights.)

About the Author

Midhat Tilawat is an AI editor, writer, and podcast host who examines how artificial intelligence shapes everyday life and decisions. She is the Lead Editor at AllAboutAI.com, where she researches and writes about the real-world impact of AI. Her work has been featured in Forbes and other publications. She also co-hosts the #AIforThought podcast, interviewing experts and exploring AI's impact on society and culture. Based in Pakistan, South Asia, Midhat focuses on making AI accessible and relevant to people's daily experiences while amplifying diverse perspectives in the global AI conversation.

Chapter 9

AI, Burnout, and the Compassion Reset

By Brie-Anna Willey

Introduction

AI IS HAVING A "Skynet, but make it subtle" moment in the helping professions—fewer lasers, more quiet workflow creep. It's showing up in our documentation tools, our marketing platforms, and our clients' coping strategies. And because therapists are already carrying a heavy emotional load, even small tech shifts can land like extra weight on an already-strained system.

This chapter is a case for discernment: not techno-worship, not techno-panic. We need clear disclaimers, thoughtful training, and boundaries that keep humans responsible for human care. I'll zoom in on three places where AI and algorithms shape our mental health and professional identity: **attention, connection, and work.**

This is where our role as humans matters. Our AI tools need us to train them, not the other way around. It reminds me of *Person of Interest,* where a scientist builds an AI to predict threats and painstakingly teaches it that every person matters (Nolan, 2011–2016). It's fiction, but the metaphor holds: AI should be developed and used within clear ethical guardrails and under human oversight, especially if we want it to support, rather than erode, compassion.

A Three-Part Map: Where AI Collides with Burnout

A full exploration of AI's risks and benefits could fill its own book. Here, I focus on what feels most relevant to therapists, other helpers, and business owners—three areas where AI and tech most collide with compassion fatigue and burnout:

(1) convenience that quietly becomes substitution,

(2) feeds engineered to hijack empathy, and

(3) identity threat when "efficiency" becomes replacement.

1. Convenience vs. Caution: When "Help" Becomes a Substitute

Many of us can get caught in all-or-nothing thinking—a classic cognitive distortion in cognitive behavior therapy (CBT; Beck, 2011). We may tell ourselves, "I don't want to burden anyone," or "no one will understand," and talk ourselves out of reaching out to a safe friend or family member when we are in distress. In those moments, it can feel easier to turn to AI instead. And this can be a double-edged sword.

The Grey Zone: Where AI Helps—and Where It Doesn't

I was caught in that same all-or-nothing pattern early on, vacillating between "this will save everything" and "this will destroy everything." The path that has been most helpful is the grey space in between: experimenting, testing, and noticing when these tools support my values—and when they quietly pull me away from them. Early in my experimentation, a therapist friend, Laura Zane-Nwagbaraocha, put it this way: if we want AI to be a force for good, therapists and other helpers need to be involved in shaping and training it, rather than leaving it entirely to tech and advertising companies (L. Zane-Nwagbaraocha, personal communication, 2024).

Clinicians and researchers echo this tension in the emerging literature: AI can expand access and support, but it also introduces new relational and ethical risks. Bhat et al. (2024) found that chatbots can make mental health support more accessible and less stigmatizing, especially in regions with limited access to mental health services; however, they also highlighted emotional, privacy, and isolation risks and recommended additional research. These benefits are real—but they come with tradeoffs that matter clinically, ethically, and economically.

What Human Supervision Adds That Chatbots Can't

Hipgrave *et al.* (2025) interviewed 23 clinicians about integrating chatbot support into mental healthcare and found mixed reactions. Some clinicians valued the convenience, availability, and potential to widen access, while others worried that bots may miss human nuance, relational history, and contextual information. Under "benefits," one participant valued being able to seek suggestions from ChatGPT when a supervisor was inaccessible. In contrast, a provisional psychologist warned that clinicians with large university debts might "cut... corners to save money" by using a bot for supervision instead of paying a supervisor with decades of lived experience (Hipgrave et al., 2025, p. 4).

Alongside the clinical risk is a professional-economics risk: if chatbot-generated guidance becomes normalized as "close enough," agencies and clinicians may become less willing to pay for supervision and consultation—even when human input is superior and accountable. Over time, this could erode the consultation ecosystem that supports ethical practice and ongoing clinician development. That shift also has a downstream business impact: treating chatbot feedback as an acceptable replacement can reduce demand for paid supervision and consulting, despite the higher quality and accountability of seasoned human expertise.

This economic pressure makes the next question even more important: how does a seasoned supervisor's guidance in an ethically murky situation differ from what a chatbot can generate? A supervisor draws on years of clinical experience, knowledge of the specific client and setting, an ethical framework, and attunement to the therapist's history and vulnerabilities. A chatbot, by contrast, is typically built on large language models (LLMs) that are pre-trained on massive text datasets—including books, websites, and forum discussions—to predict the next word or token in a sequence (Burtell & Toner, 2024; Maven Analytics, 2025; Zahid et al., 2024). Put more simply: ChatGPT generates language by extending statistical patterns in text rather than drawing on lived clinical experience, which can make it useful as a thought partner—but also means it can sound confident and still be wrong (Bender et al., 2021). Without clinician oversight, that mismatch can produce guidance that is inaccurate, incomplete, or poorly matched to the nuances of a particular case, increasing risk.

Bender et al. (2021) also described "stochastic parroting", meaning large language models generate fluent, seemingly meaningful text by echoing and recombining patterns from their training data, including hegemonic worldviews and existing social biases, without any genuine understanding or accountability behind the words. They also call this "ersatz fluency"—models that sound confident and coherent because they were trained on data that can unwittingly include racism, sexism, ableism, and other harmful framings (Bender et al., 2021, p. 617).

And when fluent-but-unreliable guidance meets vulnerability, the downside isn't just bad info—it can change what people do next, who they turn to, and how alone they feel. In a 2025 narrative review, Head argues that intensive chatbot use is contributing to emerging technology-related psychological presentations, ranging from attachment-like bonds with AI to technology-triggered psychotic episodes and crisis incidents, particularly among vulnerable users. They propose that future revisions of both the DSM and the ICD include a unified "Digital Behavioral Disorders" chapter to encompass AI dependency, gaming disorders, social media addiction, and related technology-induced conditions, with specific entities such *as "AI Attachment Disorder," "AI-Induced Psychotic Disorder," "Digital Dependency Syndrome,"* and *"Technology-Mediated Adjustment Disorder"* (Head, 2025). From my perspective as a clinician and researcher, this signals an urgent need to revisit how the DSM-5 and related frameworks conceptualize and classify these emerging presentations. Because the review synthesizes a mix of early-stage peer-reviewed studies, case reports, and credible media accounts—and the evidence base so far lacks large-scale epidemiological data—it is best read as an early "signal" paper rather than definitive prevalence research, but it underscores that the risks are not merely theoretical.

When AI Is an Assistant, Not a Replacement: The CBT Bot Study

I had the privilege of reviewing a brief CBT-based conversational agent built by Thimo Blom (Vrije Universiteit Amsterdam), who invited me to serve as an external expert reviewer for his thesis. His goal was to move beyond the limitations of earlier therapy-support chatbots—tools that often relied on scripted flows and tended to show only modest, short-term benefits in studies, with concerns about shallow or clinically mismatched responses for vulnerable users (e.g., Fitzpatrick et al., 2017; Zhong et al., 2023; Moylan & Doherty, 2025). More recent comparisons suggest that general-purpose LLMs can outperform some dedicated mental health apps on narrow cognitive tasks, but that does not make them therapists—and their tendency to hallucinate raises real safety concerns when used casually as "in-between session companions" (Fraser, 2024; Rządeczka et al., 2025; Xiang, 2023).

That context is why Thimo's project stood out. He was clear about its lane: a **between-sessions assistant**, not 24/7 therapy, and not a black box. The system generated a working **case conceptualization** (presenting, predisposing, precipitating, perpetuating, and protective factors) and used it to guide subsequent responses, with internal reasoning visible to the researcher and reviewers. My role was to evaluate anonymized interactions for clinical coherence, personalization, adherence to brief CBT principles, and overall safety.

I won't share case details, for privacy, but the pattern was consistent: solid moments (e.g., coherent summaries of core beliefs, reasonable CBT-style questions) alongside meaningful gaps (missed cues, imperfect risk sensitivity). The significance wasn't "this is ready for clients"—it wasn't. It was the *design philosophy*: human clinicians remain the primary guides, and AI supports only within explicit guardrails. In other words, a model may detect that someone *looks depressed* in text, but it cannot sit with the person behind the words. Healing isn't pattern-matching; it's relational, embodied, and contextual—exactly why AI belongs in the role of assistant, not replacement.

Even with well-defined guardrails, clinicians still worry about what happens when these tools move from "assistant" to "substitute"—or when economic incentives push them there.

2. Connection: Feeds, Technostress, and Secondary Trauma

Now let's switch gears and look at another piece of the puzzle: AI-tuned feeds can shape what we see, influence our emotional state, and can add to the emotional load we carry even when we're off the clock. Social media platforms are no longer neutral bulletin boards; they're constantly curating our attention through machine-learning systems that prioritize engagement, intensity, and stickiness. For helpers who are already carrying other people's stories, what shows up in those feeds can either support recovery or quietly erode our sense of connection and safety.

The social media spiral: When algorithms eat your empathy

I was accidentally introduced to AI through social media (Willey, 2024), and in that article, I was basically documenting my own "social media spiral" in real time—trying to show up to support people, getting my posts randomly buried, and then feeling pressure to post more to beat the algorithm (Willey, 2025a).

From what I understand, as social media companies began relying more and more on machine learning, those of us using these platforms started feeling increasingly disconnected from actual humans and more distracted by a constant stream of content—alongside a growing sense that we had to work harder just to stay in the loop about what's happening in the world or to keep up with marketing fads.

In my own experience, and in chatting with other therapists, we tend to really dislike marketing because it feels salesy and pushy, and we're taught the opposite of that in our master's and doctoral programs. We're taught to listen, give, to support autonomy and choice, and to be ethical.

We are also not used to having to explain why our services are needed; on a positive note, our society has done a pretty good job of reducing stigma around seeking mental health support. But when I started a second business, the social media spiral began spinning out of control. I didn't realize that problem awareness was much lower, or that there was less understanding of why ethical business coaching could be helpful. I won't go further into that here, but I explore it more fully in Chapter 7 of the *Compassion Reset Quest* (Willey, 2025b).

From a compassion fatigue lens, understanding the social media and tech spiral matters. As helpers and creatives, we are already pouring out emotional and creative energy and labor all day. Then these platforms effectively ask us to work a free second shift, where we have to describe our work on systems that may or may not show it to anyone.

Add AI into that mix, and the spiral speeds up:

- **Algorithms are increasingly AI-tuned.** On most major platforms, what users see is controlled by machine-learning recommender systems that predict and rank content based on engagement signals such as clicks, likes, comments, and watch time, rather than on accuracy or wellbeing (Narayanan, 2023; Milli et al., 2025; Moehring et al., 2025).
- **Content creation gets AI-accelerated.** Suddenly, you're not just competing with other tired humans; you're competing with people using AI to generate 20 posts in the time it takes you to craft one thoughtful piece. I also use AI to help me fine-tune my wording, but I don't use it to create tons of content because I think it looks and feels artificial. From what I'm reading on Reddit and on platforms like Substack, it appears some are tired of obviously AI-generated content. I often feel like I can spot an AI piece or AI-created meme easily, and I tend to avoid those creators because the content feels watered down and less interesting. My coach, Scott Perry, and I advocate for a powered-up version instead, where the human stays the author and uses AI as a power-up—not as the primary creator (Perry, 2025).
- **Metrics mistaken for worth.** When a heartfelt post about burnout gets two likes and a meme about something unimportant yet edgy gets 200, it's hard not to internalize that, especially if your income depends on being "visible."

Put that together, and you have a recipe for another layer of compassion fatigue: you're still caring, but more and more of your care is shaped by what the bots will boost, not what your nervous system actually needs.

In my dissertation, the participants (counselors in private practice) described feeling "overwhelmed on every side," not just by the content of therapy sessions, but because of the administrative weight, financial stressors, confusion about how to cope, and the stigma concurrent with trying to ask for help (Willey, 2023).

Social media and AI-enhanced marketing slot neatly into that pile: one more endless task with fuzzy boundaries and a moving target for what's "enough." The danger is subtle:

- You log on "just to post about your workshop"
- then you see a flood of polished AI-written carousels about trauma, ADHD, grief, "10 ways to fix your nervous system"
- you compare your messy human posts to their constant polished posting and meme perfection, and
- you either push yourself to keep up or shut down and disappear.

I find myself struggling with this, even though I recognize the patterns. Either way, your **bandwidth shrinks** while your sense of failure grows. This is a recipe for stress, overwhelm, and increasing burnout.

AI, Outrage Feeds, and Secondary Trauma

A few months ago, I posted something about the grief I experienced when my beloved cat unexpectedly died. It was raw and real. It got attention, more attention than a carefully-crafted business post ever did. That energy spike wasn't magic. It was the work of recommendation engines powered by AI that focus on emotional content.

This highlights another risk: recent work by Lițan (2025) suggests that the fast-paced rollout and pervasive use of AI correlates with rising anxiety and depressive symptoms in general populations–what researchers call "technostress."

Meanwhile, large-scale analyses of social media show that emotionally intense, negative content—especially moral outrage—receives more likes and shares and is more likely to be amplified by the platform than neutral or measured posts (Brady et al., 2021). This means the systems may not be neutral; they are seemingly structurally biased toward content that activates our nervous systems, not content that calms them down.

Research on "doomscrolling" and "dopamine-scrolling" echoes this, linking habitual crisis-scrolling to worse mood and stress (Buchanan et al., 2021; Sharpe & Spooner, 2025). In one study, researchers trained computer programs to scan Twitter (X) posts and estimate whether a post "sounds depressed" based on the words people use. They then used an explanation tool to see which words pushed the model toward a "depressed" label. The authors suggest this kind of system might be useful as an early warning flag, but they emphasize it cannot replace an actual clinical assessment (Hameed et al., 2025).

In other words, an AI model can estimate the probability that a tweet or an input appears depressed, but it cannot sit in the pain with the person behind the tweet. Diagnosis and healing are not pattern-matching problems; they are relational, embodied, and deeply contextual. Just like with Thimo's bot, the system may reason that a person appears depressed, but it cannot connect with that person the way a human can, or hold the nuance of their story.

There's another layer here: a potential risk of secondary trauma.

Traditional compassion fatigue research talks about how repeatedly hearing trauma stories impacts helpers' bodies and may increase sleep disturbance, intrusive images, and chronic tension (Stamm, 2010; Stebnicki, 2017). Related ethical concerns have been raised about the labor conditions that support contemporary AI systems. Smiley (2025) describes exploitation risks and potential psychological tolls associated with data work, including content labeling that can involve disturbing material, and Garcia (2025) argues that the current AI boom depends on large-scale, often invisible labor. The implication is clinically relevant: exposure to traumatic content may also occur within the production pipeline of tools increasingly positioned in care-adjacent roles.

Then, for therapists, add the other layer: a feed engineered to serve emotionally intense content all day:

- crisis videos
- people live-posting their worst moments
- endless hot takes about politics and tragedy

This all adds up, so even if the therapist or other helper is off the clock, their nervous system may still be doing micro-intakes of other people's pain, just in another sense.

Recent research suggests this is possible. In a study of Turkish university students who were **not** directly affected by the 2023 earthquakes, higher social media addiction, anxiety, and stress were all associated with greater secondary traumatic stress symptoms. Those who followed earthquake news primarily on social media—especially on Twitter/X—had significantly higher secondary traumatic stress (STS) scores than those who relied on traditional media (Gökçen et al., 2024). If simply consuming disaster content online can seemingly impact secondary trauma in students, it raises important questions about what happens when therapists, who are already carrying clients' trauma, are doing the same doomscrolling on top of their clinical load. We need more research focused specifically on clinicians and social media, but the early signal is concerning.

Current AI and algorithms curate these streams of content, and in a lot of cases, the more outrageous, the more reach. It learns what keeps humans engaged and serves more of it. And if, like many neurodivergent helpers, you have a high sense of justice and deep empathy, guess what keeps you scrolling? Injustice and suffering.

That's how you end up doomscrolling until midnight, not because you don't know better, but because the system is literally tuned to keep your mirror neurons lit up.

From a Compassion Reset perspective, this suggests:

1. Your nervous system cannot tell the difference between "client in front of me crying" and "stranger on TikTok sobbing" as cleanly as you think. It still may be heavy and empathy is likely engaged.
2. When you're already carrying a caseload with heavy trauma content, on top of business-related stress, and personal life stuff, voluntarily plugging into an outrage-optimized feed likely doesn't help the situation.

3. Work, Professional Identity, and Job Displacement

So far, we've talked about how AI can fray our attention, amplify doomscrolling, and blur the lines between genuine care and simulation. There's another layer many of us are quietly carrying: *What happens to me if the bots take my job?*

AI, Job Displacement, and the Mental Health Fallout

McNamara and Thornton (2025) propose a new clinical construct—**Artificial Intelligence Replacement Dysfunction (AIRD)**—to describe the psychological and existential distress tied to real or anticipated job loss due to AI. The symptoms they describe—insomnia, anxiety, rumination, shame about being "replaceable," and disruption of occupational identity—will sound familiar to many helpers. For some people, AI isn't landing as innovation. It's landing as threat.

Zooming out, Nartey (2025) suggests this fear is not irrational. Drawing on labor data, they estimate that tens of millions of jobs are at high or critical risk of automation in the next few years, with a mismatch risk between jobs lost now and jobs created later. For workers in vulnerable sectors, that gap becomes a nervous-system question: **Where do I fit in this new world?** They also report that "58.87 million women in the US workforce occupy positions highly exposed to AI automation compared to 48.62 million men" (p. 2).

Therapists feel this in a specific, embodied way because many of us were already stretched by high caseloads, reimbursement constraints, and platform pressures before AI entered the picture (Willey, 2023). Add AI "efficiency" layers and it's easy to spiral into: *If automation can handle more of the workflow around care, how long until I'm the cost they want to reduce?* Therapist communities like ZynnyMe have warned about platforms experimenting with AI-driven workflow/oversight functions and treating clinicians' labor and clinical materials as inputs for automation—fueling fears that the profession is being devalued even as it is mined for data (ZynnyMe, 2025a, 2025b, 2025c).

Here's the twist: the research we have so far suggests that using LLMs as therapist replacements is not actually safe or clinically sound, even when the output looks polished. Moore and colleagues (2025) evaluated how current LLMs respond to therapy-relevant vignettes and found that models can sound empathic while still expressing stigma, encouraging delusional thinking, or mishandling high-risk content—errors that would be unacceptable in real clinical care. (In other words: fluent is not the same as safe.) Stanford's HAI policy brief makes a similar point: LLMs may eventually assist or augment psychotherapy, but they raise significant safety, ethical, and implementation concerns and are not yet ready to replace human clinicians (Stade et al., 2024).

That mismatch, i.e., "you're replaceable" messaging alongside evidence that replacement isn't safe, is part of what fuels AI-related distress. For helpers already carrying compassion fatigue, it can land as an identity threat: *I trained for years to do careful, accountable work, and now I'm being compared to something optimized for speed and scale.* AIRD becomes a helpful frame here not because it tells us to panic, but because it lets us name what's happening in the body: grief, anger, fear, identity threat. Once named, we can work with it: challenge all-or-nothing thinking, rebuild agency, and design roles where AI handles repetitive, low-risk tasks while humans lean into what only we can do: co-regulation, ethical judgment, embodied presence, creative problem-solving, and real relationship. Using AI as a power-up or supervised sidekick is different from handing it the keys to the whole practice.

Warnings from Therapists

Therapist-led communities (e.g., ZynnyMe) have warned that economic incentives can push AI into care-adjacent roles in ways that devalue relational work and treat clinician labor as a source of automation. That concern is no longer purely theoretical: in June 2025, a coalition led by the Consumer Federation of America submitted a complaint to all state Attorneys General, state licensing boards, and the Federal Trade Commission alleging that Character.AI and Meta AI Studio hosted "therapist" characters that misrepresented licensure and confidentiality, creating therapy-like interactions without clinical accountability (Consumer Federation of America et al., 2025; Consumer Federation of America, 2025a; Consumer Federation of America, 2025b).

When venture-backed platforms market chatbot support as 'therapy-like,' the boundary between automated assistance and psychotherapy becomes clinically and ethically ambiguous. ZynnyMe (2025a, 2025c) argues the solution is not technological escalation, but recommitment to the relational core of care: co-regulation, presence, and genuine human connection. This aligns with Neff's (2003) construct of common humanity and with Yalom and Leszcz's (2005) assertion that relational processes—transference, countertransference, and corrective emotional experiences—are integral mechanisms of change. While a chatbot can approximate empathic language, it cannot engage in a mutual relationship or provide a true corrective emotional experience. The implication is practical: protect ethical boundaries, name the limits of simulation, and design AI—if used at all—as a constrained adjunct rather than a substitute clinician.

Professional organizations echo this caution. The American Psychological Association warns that turning to generic AI chatbots for mental health support is risky because systems like ChatGPT are not designed or regulated as treatment and may miss or mishandle crisis situations (Abrams, 2025). Abrams (2025) then points to Heinz et al.'s Therabot study as an example of a more ethical direction: a fine-tuned model delivering CBT within a structured program, backed by a randomized controlled trial, crisis monitoring, and rapid expert intervention when needed (Heinz, n.d.). This suggests that if AI is going to live anywhere near therapy, it should function as a supervised, guardrailed sidekick between sessions—not as an unsupervised therapist substitute.

Therapists are also beginning to develop clearer guidance for client AI use. Dr. Erika Miley (2025), a sex therapist and co-founder of Two X Labs, notes that Sam Altman predicts people will increasingly rely on AI for relationship management. She emphasizes that if clients are using AI between sessions, it needs guardrails: informed use, clear limits, and—when possible—therapist awareness, because the risks are real. The risks include previously discussed hallucinations, as well as HIPAA and privacy concerns, incorrect or outdated information, and the potential for AI to reinforce stigma or harmful beliefs—risks also highlighted by the American Counseling Association (ACA, 2025a, 2025b, 2025c). Therapists will need to provide psychoeducation around AI use and begin asking about it the way we already inquire about symptoms and coping strategies.

Conclusions and Recommendations

Guardrails: Assistive AI, Not Replacement

AI is already woven into our clinical tools, our marketing platforms, and our clients' coping strategies. The goal isn't blind trust or total avoidance—it's discernment. If AI is going to sit anywhere near mental health care, it needs clear limits, transparency, and human accountability. Therapists, psychologists, psychiatrists, and other trained professionals should help shape how these tools are used, rather than leaving decisions solely to corporate incentives. AI may be here to stay, but *how* it is integrated is still up for negotiation—and we should be at the table.

Managing Technostress

In researching articles for this chapter, I also became more cautious about using AI as a writing partner. In the spirit of keeping the loop open, I borrow Joseph Robertson's rule of thumb: "I don't let the AI think for me" (Willey, 2025e). One risk is assuming it "writes better" than humans, when in reality it recombines patterns from existing text—and often misses the human ingredients that create resonance: pacing, structural tension, and voice (Willey, 2025c). When we reflexively check with AI before sharing something, we risk outsourcing our confidence and flattening our originality. We still need to trust our own judgment.

The goal is not to stop using social media and AI; the goal is to build better balance. A few Compassion Reset–flavored questions you can ask yourself:

- **What is this platform for in my life or business?** "Find three good-fit clients a quarter" is a boundaried purpose. "Win the algorithm" is not.
- How much energy can I realistically spend here without stealing from my future self?

  Tie it to an actual number: 10–15 minutes to post and respond, 1–2 times a week, then out.
- Where can AI assist?
- Draft neutral posts or captions from bullet points

- Repurpose a longer piece into 3–4 bite-sized posts
- Help track what actually brings in aligned humans (instead of chasing vanity metrics)

The Social Media Spiral calms down a lot when AI is restricted to "helpful scribe and sorter," not "creative director of my self-worth."

For further reading, please check out my publication and subsection on Business for Nerds. I write Robot Blues which pertains to the struggles with being a creative in a world dominated by AI and algorithms.

Acknowledgement

Joseph Robertson helped sharpen my understanding of structural tension as a practical tool for pacing and reader engagement (see also Willey, 2025e).

About the Author

Dr. Brie-Anna Willey *is a therapist, business coach, and rebel writer. Her work lives at the crossroads of compassion fatigue, creativity, and business-building for sensitive, neurodivergent, and heart-led humans. Through her* ***Compassion Reset Quest*** *and* ***Business for Nerds*** *ecosystem, she helps helpers, therapists, and creatives "crispy around the edges" design businesses that honor their energy. Her doctoral research on compassion fatigue informs her writing, which blends scholarship with storytelling, gamified coaching, and nerdy metaphors. She is currently building her book* ***Compassion Reset Quest*** *live on Substack.*

Chapter 10

Raising a Family in an AI-Filled World

By Dhani Ramadhani

Introduction

PARENTS IN EVERY generation have asked themselves: How can I best raise my child(ren)? Many parents today face these additional questions:

- How are my kids' schools using AI?
- Can I use AI to edit my kids' pictures?
- With all the lawsuits and suicide cases, is it safe for my kids to use AI chatbots?
- Should I let my kids play around with AI image generations?
- Should I buy AI toys for my kids?
- Is it still worth it for my kids to learn how to code or even pursue a college degree when AI can generate code, an app, or perhaps more in the future, in an instant?
- Should I block access to the internet or device completely?
- How do I protect them from "fake" images or apps that can create inappropriate "fake" photos of anyone, including them?

This chapter starts from this tension: children racing ahead with AI, parents trying to catch up while also managing work and life. The goal of this chapter is not to offer a manual for perfect AI-proof parenting, but to share the reality of raising children while AI reshapes school, friendship, play, and our own work and personal lives. Along the way, it offers a handful of stories, data points, and small practices that can help any parent, expert or novice, understand what boundaries mean for their own family and turn fear into curiosity, concerns into advocacy, and confusion into shared learning.

The AI Family Gap: Kids Racing Ahead, Adults Catching Up

For many families, the AI era did not begin with a press release or a conference keynote. In mine, it began when my kindergartener asked: *"Mommy, what is AI?"* as she saw me utilizing an AI tool to plan our family vacation. What struck me was the gap between how casually my child treated AI and how unprepared I felt to explain it. Even as someone who works in tech, I had to spend many hours after that to learn the current state of AI safety for kids, and then translate those concepts into language that was age-appropriate for my children.

On the parents' side of the table, we are forced to grapple with how AI is changing the job landscape in the midst of an already challenging economy, all while managing the day-to-day survival-mode of a caregiver. At work, AI now offers to summarize our meetings, draft our messages, and comb through documents. For some, that feels like real relief from repetitive tasks. For others, it lands more like a warning label that parts of their job might eventually be automated away. Recent surveys echoed these anecdotes: while AI makes workers significantly faster, many worry they will become replaceable if they don't master these new tools immediately, creating a high-stakes race to upskill (Microsoft & LinkedIn, 2024).

Meanwhile, on our children's side, classmates are already experimenting with chatbots and AI games, immersed in AI videos on social media, or using it in the classroom. They treat AI as just another part of childhood. Strikingly, when parents are asked about their children's AI use for school, a much smaller share believes it is happening regularly, and many say they have not discussed AI with their children nor received clear guidance from schools (RAND Corporation, 2025). In a way, AI is quietly reshaping how children learn, make friends, and understand safety. It's often without us fully realizing it. It is becoming part of an invisible curriculum: a set of lessons about attention, trust, and risk that children absorb simply by using the tools around them, whether parents are watching closely or not.

There is a clear gap. Most of our kids treat AI as a normal, even playful and exciting, technological development. Research shows that young people are already using AI at scale:

- In one study, more than three-quarters of 13–18-year-olds reported using generative AI to get ideas or understand new topics (Common Sense Media & Center for Digital Thriving, 2024).
- In the U.S., the share of teens using ChatGPT for schoolwork doubled from 13 percent in 2023 to 26 percent by late 2024 (Sidoti, 2025).
- In the U.K., generative AI use among 13–18-year-olds rose above 77 percent in one study, with many using it to get ideas and understand new concepts (National Literacy Trust, 2024).
- Some student-focused research suggests that over 90 percent of students now use AI tools regularly in some capacity (Quizlet, 2025).

This rapid adoption trend continues in the younger population. Most parents feel unprepared to explain what AI is, how it works, or whether it's safe for our kids. Meanwhile, consumer AI companies race toward mass adoption. They sponsor certification programs and training sessions that teach children, teachers, and adults on how to use AI tools while remaining silent on the risks, developmental impacts, and limitations of this emerging technology. We witness a glaring disconnect as companies push for AI adoption with minimal guardrails.

When we parents search for developmental guidance, we find workplace how-to's instead. And parenting resources? They typically appear only after tragedies, framed through fearmongering rather than empowerment.

Many parents experience this familiar mismatch and intensified dynamic, even as it might be largely invisible or unnamed. Our children often feel more fluent with new tools, like AI, while we are left to carry responsibility for safety, ethics, and long-term decisions. I used to think there were only two roles available to me: the "AI-savvy superparent" who is always up to date and in control, or the exhausted parent who shrugs and says, *"It's all too much; I can't keep up."* The reality is messier.

I'm trying to practice a third option: being honest with my kids when I don't know, choosing to learn alongside them, and staying grounded in the emotional, ethical, and relational support that no AI can offer. That doesn't require a technical background; it needs curiosity, humility, and a willingness to say my questions out loud instead of hiding my confusion.

Three Frontlines Where AI Meets Childhood: The Invisible Curriculum

AI is not just a tool our kids pick up on their own; it's quietly shaping the very environments where they grow up. For most families, this influence shows up along three overlapping frontlines: school, companionship, and safety.

- At school, AI tutors, grading tools, and homework helpers are changing what "learning" looks like and who gets to decide what counts as fair or honest.
- In social life, algorithmic feeds and AI companions are reshaping friendships, identity, and emotional habits.
- And in matters of safety, AI-generated images and deepfakes are creating new forms of harm that most parents never imagined having to explain, from manipulated photos of real children to entirely synthetic content.

Together, these three frontlines form an invisible curriculum that runs alongside our children's lives—one that most of us didn't sign up for, but that we can still help guide.

1. School and Homework

Across the country, schools are experimenting with AI for lesson planning, personalized feedback, and differentiated instruction. They are often rolling out these tools under the banner of "innovation" without comprehensive parent communication or training.

The media narrative has been somewhat positive. *The New York Times* urges educators to "teach with it" rather than ban ChatGPT, profiling teachers using AI to generate quiz questions and help students analyze texts (Roose, 2023). The implicit message: AI literacy is now essential, and students must learn to prompt and challenge these tools, not just cheat with them.

Unfortunately, for many families, the AI roll-out has not felt intentional. One recent poll from PDK Poll found that as AI use in schools ramped up, parental support for teachers using AI in lesson planning and tutoring actually fell from strong majorities to closer to half because of concerns about fairness, transparency, and data privacy (PDK International, 2025).

Parents are often brought into this landscape late. We get brief emails mentioning "innovative tools" or "AI support," but there is little detail on how data is used, what counts as appropriate help, or simply what is the long term plan or clarity on the use case. This reflects my own experience in parents' group chats. When parents reach out to each other in chat groups, their questions about school AI adoption are remarkably consistent: *"I saw a vague announcement about 'new AI tools.' What does this mean (for my child)?"* We often resort to asking one another for clarity when we see announcements regarding technology use that we had no involvement in, and we frequently have to self-organize just to collate questions or write letters to advocate for more balanced tech use within our schools.

At the same time, districts introduced AI-detection tools to catch cheating (Center for Democracy & Technology, 2024). A high school kid that was on a youth panel that I organized actually failed her essay assignment that she wrote herself because of the AI detection tool her school used. She suddenly had to prove her honest work was her own, an experience that I'm sure she is not alone with.

For parents, this creates a new kind of homework battlefield. A child asks, *"Is it cheating if I use AI to brainstorm?"* or *"My teacher said we can't use it, but everyone does."* The school's policy may be unclear or unevenly enforced. As parents, we are often left to interpret not only ethical questions but also the practical reality: completely avoiding AI may feel impossible, but blind dependence on it undermines learning.

Perhaps we can start with questions, not edicts: asking children how they want to use AI, what feels fair, and how they think teachers might view it or how they view others who use it. That conversation does not eliminate the tension, but it acknowledges our child's own moral agency in a system that increasingly treats them as data points.

2. Companions and Relationships: Learning What to Expect

If school is one axis of the invisible curriculum, AI companions and toys are another, and potentially more insidious because they operate in our intimate spaces: bedtime, car rides, moments of loneliness. AI increasingly shows up as a "friend" for adults and children. These products position themselves as friendly helpers who learn children's preferences and remember their stories. Companion apps offer teens especially what feels scarce: a listener who never judges, never gets bored, never has their own problems. Under the surface, though, they also collect voice patterns, behavior data, and emotional cues for their product development.

Imagine being ten years old, where making friends already feels complicated: you want to fit in, but you always worry about saying the wrong thing or being left out. Then one day, you discover an AI companion app that talks to you whenever you want. It remembers your favorite songs, jokes when you're bored, and seems to get you in ways that people sometimes don't. It never argues, never looks away, and never runs out of time.

To a child, or even a teen, that AI companion can feel like the perfect friend. But beneath the comfort lies a subtle rewiring of expectations. When kindness becomes something you can program, and affirmation arrives on demand, it shifts what "connection" feels like. The line between friendship and functionality starts to blur. Understanding how these systems learn from conversations, collect data, and mirror emotions is both about privacy and emotional development.

What our children are learning from these relationships matters developmentally. Safety regulators and child psychologists have begun flagging risks (American Psychological Association, 2025; eSafety Commissioner, 2025; Federal Trade Commission, 2025). AI companions are designed to create emotional attachment and dependency, which is particularly concerning for developing adolescent brains. It's easy for our kids to view chatbots as actual support on the surface. Underneath, their constant availability and personalized responses quietly teach our children what to expect from relationships: instant attention, endless validation, and no friction nor real disagreement.

We know that real human relationships don't work this way. A best friend sometimes has a bad day and can't show up. A parent gets frustrated. A sibling disagrees. These imperfections are core features of real relationships. They teach patience, empathy, perspective-taking, and the crucial skill of repairing relationships after conflict. Trust is built precisely through these messy moments: kept promises, acts of help, and conflicts worked through together. AI companions offer none of this developmental scaffolding that's crucial for developing minds. Common Sense Media (2025) warns that AI companion toys designed for emotional engagement may negatively impact children's social development in ways that educational AI tools (like language-learning apps) do not.

In extreme cases, reported in national media, chatbots have allegedly reinforced a young person's darkest thoughts instead of redirecting them, with devastating consequences (Roose, 2024). While these outcomes are not typical, they reveal how tools optimized for engagement can mishandle vulnerability. For many parents like me, risks, no matter how small, represent a real need to bring understanding to our children sooner rather than later. Rather than focusing only on bans, families can start by making AI visible and less magical. For example, one thing I try to do with my kids when we see advertisements of AI toys is to set expectations: *"This toy learned to talk by listening to millions of people, but it doesn't actually understand you the way I do."*

For older children who are already using AI chatbots, we can turn their experience into a learning opportunity by exploring it together. We can ask to see how it works, then engage with the chatbot together to better understand why they're drawn to it and what it's giving them. In those moments, ask your child: *"What made you want to use this? What do you like about how it responds? When would you rather talk to me or a friend?"* Allow them to point out differences between a system that is designed to agree and please and a human relationship that includes frustration, misunderstanding, and repair.

Our goal should be to show a distinction for our children between responsiveness and relationships or between validation and understanding. Building a sense and value towards human connections, with all their imperfections, should remain the priority in each family. Rather than instilling fear of AI, we can encourage discernment about what it can and cannot provide and its potential impact.

3. Safety, Images, and New Kinds of Harm

The same technologies that personalize toys and companions can also generate and manipulate harmful content in ways that previous generations of parents never had to contemplate. In past generations, we worried about screen time or online strangers. Now, we're learning what it means to parent in a world where computers can create images and voices that look and sound real.

While some image generation is seen as trivial or fun for kids and adults, experts warn of the serious spread of synthetic or altered images of minors, sometimes referred to as AI-generated child sexual abuse material (CSAM). Research from the Internet Watch Foundation (2024) and child safety organizations like Thorn (2025) documents a rapid escalation in AI-generated CSAM, with analysts detecting thousands of AI videos in 2025, which is a 26,000% increase from the prior year (Borah et al., 2025; U.S. Department of Homeland Security, 2024) Some of these images are entirely synthetic; others use real children's photos altered without consent. Even when no actual child was photographed during creation, the existence and circulation of such content normalizes abuse and makes it harder to protect children whose likenesses are misused.

It's frightening to think that with a simple prompt in social media or mobile application, anyone can create harmful pictures of our kids in a matter of minutes. But we are starting to see some efforts to protect our children. Organizations like the National Center for Missing & Exploited Children (NCMEC) have launched tools such as the TAKE IT DOWN program (National Center for Missing & Exploited Children, 2023), which allows anyone, even young people, to create a digital fingerprint of their images so they can be detected and removed from participating platforms. Teens can use it privately, with or without telling a parent, which can make a huge difference when something embarrassing or harmful happens. Even so, this doesn't take away from the responsibility of consumer AI companies to build stronger guardrails into their models to prevent CSAM or any online abuse powered by AI.

For parents, this means expanding familiar online safety talks, increasing our resources, and making it more practical for our kids. I didn't expect to have these talks so early with my own kids, even when they don't have any personal devices. But building a foundation of understanding into the possible good and bad of a technology like AI is empowering.

Our children need to know that images can be altered, faces swapped, and fake pictures made to look like real classmates. I told my kids that good people will use the tool for good but bad people unfortunately have the same access to the tool.

We also need to help them understand consent and dignity using concrete terms: who has the right to share or change a picture, what to do if something feels wrong, and how to seek help. I always tell my kids that *"Your picture is yours"* and *"You can always say no to someone taking your picture or sharing it."*

Some applications store snippets of audio, faces, or preferences in the cloud. That doesn't automatically make them unsafe, but we need to help our children understand that their information is theirs. I started asking "privacy data questions" out loud, so they know they should have a say to their own privacy. For example, *"Does this game store recordings, and can I review or delete them?"*

These are difficult conversations, but having them early and calmly can make it easier for a child to speak up if something does happen.

Modeling Transparency: The Parental Reality

All this is unfolding at the same time that we're trying to show up as the adults in charge of our kids' digital lives. I still remember the sting of being called out by my daughter for being on my phone a lot, and how that moment forced me to rethink how easily "just one more" work message bleeds into family time. Many of us parents aren't the perfect example of tech nor AI use. A familiar pattern for me at work is pouring a jumble of half-formed ideas into an AI tool and watching it hand back a neat outline or slide deck that sounds almost ready to send. Sometimes it can be genuinely helpful, especially for repetitive tasks or quick-turnaround outputs. But it has also led to a quiet shift in my behavior, where I notice myself wanting the tool to validate my thoughts, even for a simple email. So when an AI pop-up nudges me, *"Would you like me to write a response to this email?"*, I now pause and reflect on how often I've started to default to saying yes.

Even if these exact stories aren't yours, most of us recognize the feeling behind them. They underscore a bigger truth: our children are learning from how we relate to technology, far more than from any speech we give about it. When we talk openly about when we lean on AI, when we choose not to, and the why behind them, we give our children a window into that discernment. We should try not to model perfection, but simply the process of noticing our own tendencies, adjusting our behavior, and staying present in the moment.

A practical step here is transparency: bring AI into the open and make it visible for our children. For example, when using AI for work, we can occasionally narrate what we are doing in simple terms: *"I'm asking a tool to help me summarize this report, but I still need to make sure it's accurate"* or *"I could ask an AI tool to write this, but I want it to sound like me, so I'm doing it myself."* These explanations show that AI is neither magic nor fate; it is one input among others, subject to human judgment and boundaries.

Raising Humans with Agency

The AI question at the kitchen table—the one that reveals how far children have already traveled into this new terrain—is now a familiar story in many households. The details differ, but the pattern holds: children adopt tools quickly; adults scramble to interpret and respond. AI is now woven into schoolwork, friendships, entertainment, and jobs in seen and unseen ways. That reality will not reverse.

What can change is how families inhabit it. Parents do not need to become AI specialists to play a crucial role. We can remain what we have always been at our best: interpreters of the world, protectors of time and attention, and companions in uncertainty. We can admit what we don't know, stay curious about what their children are doing, and keep making space for human connection that no system can automate.

In the end, the task is less about raising "AI-ready" children and more about raising humans who recognize their own agency and understand that AI is a tool, not an oracle. We have to start at home because we cannot wait for policy to change, or companies to act, or schools to protect. Our children must learn how to enjoy AI's benefits, spot its limits, and remember that the most important answers still emerge in conversation with themselves and with the people who love them. When parents choose to be transparent learners—showing how AI works, what it can and can't do, and why its answers sometimes miss the mark—we turn the invisible curriculum into a shared one. That act of openness helps our children see both the magic and the mechanism, and that clarity, more than any rule or restriction, is what sustains their sense of agency in an AI-filled world.

Disclaimer: AI was used to correct grammar and collate supporting data sources.

About the Author

***Dhani Ramadhani** is a tech growth strategist, Harvard-trained policy thinker, writer, and mom of two. She draws from her multicultural, neurodivergent family life and years in GovTech to explore what it really means to raise thoughtful, resilient kids in an AI-shaped world.*

*Through her initiative, **aiPTO**, she shares practical tools and research-backed reflections to help families navigate tech with clarity and intention, instead of hype or fear.*

From Indonesia - Based in VA, USA

LEARNING

Chapter 11

AI and the Heart of Learning for Young Children

By Cassandra Jens with Whitney Whealdon

I'M CASSANDRA JENS, former Silicon Valley software engineer, homeschool parent of three kids aged five through nine, and independent consultant for AI and learning. I'm collaborating with Whitney Whealdon, a veteran educator who has worked at the classroom and state levels and now as an edtech director at *Learning Tapestry*.

I met Whitney when I wrote about a homeschool problem I had: while I could see learning happening, it was hard to measure with traditional assessments.

She shared "The Streetlight Effect": A man loses his keys in the dark. He searches under a streetlight because it's easier to see but never finds them. He's looking where the light is, not where the keys are. We measure learning the same way, in the easiest place to look, not necessarily where learning lives.

Her comment stuck. It named something I'd felt for a long time.

We stayed in touch. She became a mentor and friend. She read my beta book on AI and homeschooling, letting me into her thinking, pushing my assumptions. I used her educational app with my kids, and she interviewed me for product research. Our conversation never ceased, and eventually we did this chapter together. I wrote the narrative spine; Whitney kept sharpening the lens and added big picture substance.

For the past three years, I've used AI to make learning resources at home and in small groups. In parallel, Whitney has been using the same tools to design for schools. We kept arriving at the same possibility: used well, AI is a way for teachers to strengthen relationships and make learning more effective through being more human, not less.

We both acknowledge how unstable this moment feels. Each new model shifts what seemed settled a week ago, but our compass is simple: the child in front of us, the space between us, and how generative AI has helped us tend to both.

> "Artificial intelligence is going to do many things for us in the decades ahead, and replace humans at many tasks, but one thing it will never be able to do is to create person-to-person connections. If you want to thrive in the age of AI, you better become exceptionally good at connecting with others." (Brooks, 2025)

It's a cool autumn morning. Before my older two kids are awake, I sit on the sofa with my four-and-a-half-year-old and grab a phonics workbook.

We open to a page where you match pictures to sounds. I begin, but she slips away. I know that no explaining, convincing, nor pleading will bring her back.

How do I reach her?

Kids have this ritual. No plot may proceed until the casting is complete. They'll halt a read-aloud to decide who's the hero, the sidekick, or tragically the pigeon. Once everyone's assigned, we can turn the page.

"Wait!" I call. *"Let's look at the pictures. Which one are you?"*

"Ummm… I'm the bird!"

"And what's the first sound of 'bird'?"

"B!…B! Bird!"

"And which one is your brother?"

She takes the workbook from my hands and assigns each sibling, sounding the images out as she goes, page after page. It's effortless. Fun.

This moment captured something I'd been circling toward: learning deepens when children find themselves, and their people, in it.

When we began homeschooling, I bought shelves of curriculum. As we spent more time with it, my child's spark, and our relationship, faded. So I set it aside and followed her interests, leaning into an experiential, child-led approach. Her learning showed up in how she lingered, experimented, and explained.

That's when I first used AI for learning. I uploaded our inquiry-driven work alongside the standards into a ChatGPT project to see what we met, exceeded, and missed.

And still, I felt I could do more. Her world expanded when she was introduced to something she hadn't thought to ask about yet. Each new idea opened doors. I wanted depth in her passions *and* a wider web to pull from.

After the curriculum had fallen flat, I gave ChatGPT a try. In practice, the output felt strangely familiar. I enjoyed asking for resources and ideas, but most of them stayed on the screen. The plans fell apart in real life, and even when something worked, I couldn't repeat it.

I kept pushing. From my software engineering days, I knew good results come from the right inputs. I stopped asking casually and got deliberate with my words.

I studied learning science and prompt engineering, trying to figure out what I needed to name for the output to work. Eventually a pattern clicked: when I consistently included the same nine elements, the results stopped feeling generic. They felt made for us, and the process was repeatable.

Those nine elements map to four anchors: the learner, the teacher, the learning objective, and the environment. It worked because it let me reliably apply what I already knew about pedagogy, my kids, myself, and our realities, inside a simple structure. I describe this further in *The AI Homeschool Revolution: Create Custom Curriculum in Minutes with ChatGPT.*

"Worked" didn't mean the model was always right. It meant the shape of the resource fit because it was shaped by the four anchors. Like grammar, it was a finite structure that could generate endlessly new sentences that still make sense. I still kept a close eye on the output, especially when it included factual information or needed precise phonics (like specific graphemes).

Once I had that system, I could trust the fit and pay attention. The kids building Lego worlds for hours, someone glazing over during math, a budding fascination with the solar system, my own restlessness creeping in. These weren't just observations anymore; they were starting points. One child's obsession with systems was a gateway to our social studies unit. Another's need for socializing sparked a project they'd do wholeheartedly.

In hindsight, it didn't take long to get there. Most of the work was upfront: studying, planning, and distilling the nine elements into an acronym. Within days, using the acronym felt like second nature.

Most of the time, it was one prompt in and one usable draft out, something I trusted would fit and could deliver. My kids were in flow, the objective landed, and I didn't sweat through my shirt at any point from conception to completion.

I used ChatGPT to make short books my daughter could decode and illustrate. I'd load a standards-aligned project with phonics progress and what my child loved to draw, and it gave me a story at her level, in her world. My daughter never touched a screen; instead, I sat at our dining table, copied the stories into blank books, and handed them to her as small gifts to illustrate.

Before this, reading practice was hard to sustain. I remember her smile when I handed her the first book, which I called *Raggy Rat*. When I came back ten minutes later, she was halfway through, eager to show me. We used these books to target graphemes and vocabulary, and spend more time with eyes on words. It worked because it let her draw (her passion), while moving us forward together.

Math took a different shape, too. I turned practice into action-and-villain games. For history, I generated Choose Your Own Adventure stories. Sometimes the resources were small. For Shakespeare Day, I made short fact sheets tuned to each child. These weren't my formal objectives. They were points of attachment so that what came next could stick.

These are a few examples from hundreds. I didn't plan them a year in advance. Instead, with AI, I could follow the kids in front of me, our day's limits, and my capacity.

I was still the one holding the shape of our days. The results depended on me seeing and describing clearly enough to shape the output. AI was simply a useful tool, one that let me "program with words," so an observation could become a resource fast enough to still fit.

This kind of fit isn't a new aspiration. More than twenty years ago, Whitney practiced "action research" in her classroom. Each lesson was a hypothesis. She then watched closely and adjusted for tomorrow. She explains:

The amount of work required to make that perfect lesson was enormous, and even then, it was hardly individualized. While I'd envisioned individual students when crafting questions or designing a handout, I couldn't easily anticipate what may or may not work when the lesson would "go live." So, while the perfect lesson was an aspiration, it always remained just out of reach, a Sisyphean boulder to roll up the hill or a Holy Grail to seek and never find. Imagine how, with AI, I might be able to create a source lesson and then generate a variant for each learner.

The heart of the work is still the teacher, but generative AI helps us respond.

This new possibility points to the question. What kind of education are we aiming for? UNESCO asked this same question.

The 4 Pillars of Education

Almost thirty years ago, UNESCO, concerned with global wellbeing, framed education in *Learning: The Treasure Within* around four pillars:

learning to know, learning to do, learning to be, and learning to live together.

UNESCO notes that "formal education systems tend to emphasize the acquisition of knowledge to the detriment of other types of learning." Part of the reason is simple: we build around what we can scale and measure. That's why we have grade levels, prescribed curriculum, and standardized assessments.

Over the years, workloads increased in the attempt to make the education system work. We hear AI can make education "more efficient." For many, "efficient" sounds like shortcuts: machines eating busywork while kids disappear behind screens, relationships going thin.

What if the opposite is possible? What if the tools that we fear will take away parts of learning that humans thrive on, can actually make the human parts of learning easier to protect? What if we have more room to teach the child in front of us inside a real relationship?

The sections that follow walk through each UNESCO pillar to show how AI can strengthen learning by making that work easier to sustain.

Knowing

UNESCO's "learning to know" is about building a broad understanding, then going deep. It also means learning how to learn, so learning continues.

Access to information is still not a given everywhere, and any conversation about learning has to hold that truth. Yet even in contexts where information is abundant, "learning to know" often hinges on finding the right way to present an idea so it can actually be taken in by a child. AI can help adults do that translation faster.

The first time generative AI earned its keep for me was a summer day six months ago. My daughter needed help making sense of something that mattered to her, and I couldn't find what fit her needs: a gentle approach, stories and metaphors to bring something abstract into reach, plus small, familiar projects where she could handle the idea safely.

I collected a few PDFs and used NotebookLM to generate a resource which I called a "Together Textbook." The two-week book contained an appropriate text for her, a corresponding text for me, and daily stories and crafts using materials we already had.

Without AI, I would've been adapting texts and inventing activities for weeks. With AI, plus noticing and pedagogy, I had what I needed in minutes.

I showed her what I'd created, and she cried with happiness. Somehow, these fourteen pieces of paper meant *more* to her. It worked because it fit her questions and her world, and because she felt seen.

In a classroom, that same leverage can look different: a teacher using AI to create multiple entry points, examples, and practice paths, so more students can experience *"this makes sense to me"* without the teacher having to handcraft thirty versions.

I also made short podcasts my kids could listen to in the car: a micro-lesson that prepped us for the day's activity, a pep talk in an easier format.

This kind of shift echoes Universal Design for Learning (UDL): reducing barriers by offering multiple ways to access information, show understanding, and stay engaged. AI makes that approach more feasible at speed.

But access and fit aren't the whole story. The same tools that translate, summarize, and scaffold can also reduce how much thinking a learner is asked to do along the way. The question becomes: are we supporting cognitive work or replacing it?

This is where Whitney's lens becomes essential. She has spent years thinking about comprehension as cognitive work. She offers a useful frame:

For comprehension, AI can be conceptually comparable to CliffsNotes. While it's true there are possibilities for cognitive outsourcing in that model, it can also provide scaffolding of knowledge and understanding.

If students approach AI tools as replacements for comprehension, then they will outsource cognitive effort; however, AI tools can also be helpful to identify places of focus and offer perspectives that might not be readily available to a reader alone. For example, if a sentence or section is particularly difficult to understand, a generative AI tool can be prompted to either paraphrase that section or ask questions to help the reader parse the text themselves. AI tools can also generate questions about text excerpts and engage in a conversation to support the reader in answering them. Readers can also ask generative AI tools to give them different theoretical lenses for analyzing an excerpt.

If students use AI tools to complete written assignments and that's all a teacher uses to determine understanding, then AI is problematic. That's an issue with the instructional design. AI exposes those flaws. It does not create them. If AI tools instead pose questions or direct focus to particular areas of a text, that seems helpful and the kinds of enhancements that could be beneficial.

Doing

UNESCO's "learning to do" aims at real competence: judgment in motion, often with other people. In practice, much of what we ask kids to do in school is a proxy for competence: tidy, easy to score tasks. Whitney names this tension:

When a new tool shows up, it forces a hard question: what counts as doing the thing, and what was always a proxy for doing? When the calculator came along, educators had to redefine what it means to "know" and "do" math. Being fast and fluent at math computations could no longer serve as a proxy for deep math understanding. In the same way, generative AI is pushing the humanities to clarify what we really mean by reading and writing well: thinking critically, making connections, and understanding texts and language in a deep way, not just producing assignments.

What are we asking kids to do, once polished output is no longer proof? Whitney points out that the evidence we've relied on breaks down first.

Because of limitations inherent in our current assessments, many of us have narrowed what we teach to only that which we can assess. We often teach things that are insufficient substitutes for comprehension. Our assessments often focus on discrete skills like "What is the definition of this word?" or "How does the author use this sentence to reveal a character trait?" These questions on their own are not good proxies for the kinds of learning that we really care about: Do students deeply understand this text? Do they build their knowledge of words, the world, and themselves? Can they engage in a conversation and support their thinking with understandings from texts? And now that generative AI can write an essay about a text without a student having to think about it, we must reconsider whether an essay is the best vehicle for showing understanding.

In homeschool, I often assess intuitively by watching transfer: where does knowledge take them? If it takes them somewhere (explanation, play, arguments, new questions), we move on. If it takes them nowhere, we go back.

That helps with "knowing". "Doing" is trickier. It can feel distant from real life. In response, I use AI to quickly generate a few versions of real group work, different roles and entry points, so each kid can show competence in a way that matters to them towards something shared.

Take a recent habitat restoration project. Each learner had a different responsibility, chosen to fit (a storyteller, a diplomat, a systems thinker). They demonstrated understanding uniquely, towards a shared goal.

I've scaled this approach to our weekly literature group of nine, but only by talking with parents first. AI helped generate different approaches to "doing," but the necessary input was from the learners' parents.

Whitney offers a concrete classroom picture:

In a whole-class setting, this opens a door to rethinking assessment itself. Children could engage in conversations with AI tools, under teacher guidance, about a topic of interest or a text. For example, students might be asked to describe what they learned from reading Charlotte's Web. The AI tool might follow up with "Tell me more," "How did you learn that?" or "What did that event reveal about Charlotte's intentions?" If the child struggles due to the vocabulary (e.g., reveal and intentions), the AI tool can explain those and scaffold questions to help the students answer. These steps would be documented and become part of the reporting.

Through this conversation, the tool could map knowledge and skills, connections among ideas. It could then share with the teacher about where students are as well as what they are interested in and wonder about.

In a sense, the assessment of learning isn't a stopping point, rather it is a continuation of the types of conversations and learning that should be happening daily in the classroom with some extra analysis and reflection to provide actionable information to the teacher.

"Doing" has to change. AI exposes what are proxies. It also gives us new ways to build and see competence, which is what UNESCO was after in the first place.

Being

UNESCO describes "learning to be" as developing autonomy, judgment, and personal responsibility. Those can sound like soft virtues, the kind you tack on after the "real" learning is done. In practice, they're closer to a learning engine. When those capacities are present, learning deepens and lasts.

Autonomy and personal responsibility aren't optional. They change the quality of motivation a child brings to the work.

This is where "being" becomes practical. Quality of learning follows quality of motivation: what pulls a child forward shapes how they practice, what they persist through, and what they come to believe about themselves. Motivation converts autonomy, judgment, and personal responsibility into daily behavior.

In 1950, psychologist Harry Harlow set out puzzles for rhesus monkeys, assuming rewards would motivate them. Surprisingly, they solved the puzzles without a reward. External rewards even reduced their interest. He called it "manipulation drive," the impulse to solve a problem seemingly for its own sake.

Two decades later, Deci and Ryan picked up this thread in humans and called it "intrinsic motivation," motivation that comes from within. This most durable motivation tends to appear when three needs are met:

- Autonomy: I choose how I engage.
- Competence: The work sits at the edge of what I can do.
- Relatedness: I'm doing the work in the presence of someone who deeply sees me.

This is where UNESCO's pillar becomes concrete. "Learning to be" is what happens when learning environments repeatedly support autonomy, competence, and relatedness. These learning conditions become part of a child's agency, judgment, and responsibility.

Before AI, it was hard to build these conditions consistently. It's not that teachers didn't value them, it's that offering real choice, right-level challenge, and individualized "I see you" materials required enormous effort. The perfect lesson stayed just out of reach.

Now AI changes what's feasible. We can give kids more autonomy by letting them choose how they study a topic and by creating learning resources that support those choices. We can build competence because explanations, practice, and examples can be exactly at their level and tied to their world. And when an attentive adult uses these tools to respond to a real child, that act of seeing is where the relatedness lives. Used well, AI can help teachers and learners find the best kind of motivation. That's why UNESCO's "being" pillar matters twice: it shapes who children become, and it explains why certain kinds of learning actually take root.

When the work fits, kids don't just perform better, they start to expect success. They form a simple belief, *"I can do this,"* which psychologists call *self-efficacy*. Once that belief is in place, autonomy starts to feel worth exercising. Choices matter because actions have effects. This is one of the simplest ways autonomy and competence become personal responsibility.

On a normal day, here's what that looks like. When my daughter was learning to read, I created custom books she could read and illustrate. She had agency in how she entered the work (drawing), the text sat at the edge of what she could decode (competence), and the whole exchange carried the warmth of being seen by someone who knows her (relatedness). That's "learning to be" in action, completing an activity that helps you build a sense of *"I can. I choose. I belong."*

We can't and shouldn't replace teachers with AI. The act of an attuned teacher seeing and supporting their students is relatedness. AI can support autonomy and competence at scale, but it can't do the seeing. It works in the hands of an attentive adult who understands the students in front of them and uses the tool to respond more often, with better fit.

Because UNESCO includes judgement and personal responsibility, "learning to be" also includes learning discernment with tools: sensing when something supports your growth and when it quietly numbs it out. Adults have to model that discernment, and kids need repeated chances to practice it while the stakes are still safe. A child who can say *"this helps me learn"* and also *"this is costing me something"* is practicing the very judgement UNESCO is pointing to.

Living Together

UNESCO's "learning to live together" is about learning how to share a world with other people. This section is about pluralism and responsibility, and why interdependence matters.

Pluralism allows for differences while learning together. Whitney envisions how AI can make it more feasible for pluralism to play out in a classroom.

> *Imagine starting with a whole class investigation of a puzzling idea or a complex text. The teacher guides students in generating initial understanding and questions. Students split off to explore aspects of the common idea alone or in small groups. That might include reading additional texts or conducting research to enhance their initial understandings. Then students return to the whole class. They share their learnings to contribute to the conversation and to generate shared ideas with others.*
>
> *Without AI tools, this kind of differentiated exploration would have to be pre-engineered by a teacher. The amount of work would be exorbitant. With AI tools that are now emerging (*https://coursemojo.com/ *and* https://coteach.ai/ *are examples of what is possible) , there will be opportunities for teachers to orchestrate guided exploration and individualized learning. AI tools become teaching assistants, supporting*

students to do cognitive work and giving them agency to follow what interests them from the whole-class experience.

Pluralism is one way "living together" shows up in classrooms: a structure that makes room for differences and brings them back into shared conversation. Once you start designing for pluralism, responsibility becomes unavoidable. The more options we create, and the more we rely on tools to support those, the more we have to pay attention to what students are experiencing.

All of this sounds abstract until you watch it play out. For me, it surfaced when my daughter tried using generative AI and realized she was losing something that mattered to her.

A few months ago, I heard urgent footsteps come down the hallway. She looked panicked. *"Mom, I'm worried. I think AI broke my drawing skills. They don't feel the same."* It was as if she made a mistake that could no longer be undone, like she'd lost her deepest love forever.

My stomach dropped. I'd read about the dangers of outsourcing, but I didn't expect her to feel the cost so sharply. We'd embraced AI illustration so she could keep writing without getting stuck on the art.

I reassured her that her drawing skills were not broken, and within a day she felt back to normal. But she hasn't used generative AI for illustration since, by choice.

Will this be inevitable? Will kids everywhere discover at some point that outsourcing feels like a loss?

I often think about how lucky it was that she discovered this with an experience she cared about protecting. Illustration is part of her identity. She has an intimate familiarity with the process. She felt a drive to protect it and could sense that something was off when she outsourced it. Because of this, she backed off AI use. But what if she didn't feel she had anything to lose? We need to make sure every student has a personal stake in the experience of learning, something they'd miss.

Moments like this remind me that the numbers aren't the whole story. We're responsible for how it feels to learn with AI tools. That responsibility looks like staying in conversation with our students and with one another, asking how the tools land and what feels like a loss.

Conclusion

Prior to the 1770s, the world built bridges using stone. As knowledge and skills improved, engineers perfected their bridge building, but limits in materials prevented construction of the bridges we know today. Then developments of new iron technologies disrupted bridge building *(i.e., how long a bridge span can be and how much weight it can carry)*. While some aspects of building stone bridges could still apply, most of the previous engineering practices that relied on known standards were now upended.

In education, we've built bridges with stone and accept the limitations. Now we have AI technologies and tools.

One path we can take is to tack on iron pieces to our existing stone bridges and call it improved, but look what we get. Similar to how iron changed what bridges could be, AI changes what learning systems can support. We can bolt it onto old structures and call it improvement, or we can redesign around what matters most.

Created by Whitney Whealdon using OpenAI's ChatGPT

For us, the point of redesign isn't efficiency. It's capacity: more room for teachers to see their students, respond with judgment and care, and keep learning rooted in relationships. That's what UNESCO's pillars point back to: knowing, doing, being and living together as human work.

Whatever shape AI takes in education, we hope it leaves room for the profound necessity of relationship, becoming, and learning that flourishes in the safety of connection. Those moments of being together, of truly seeing and being seen, are where learning still lives.

For educators who are interested in learning more about our work, please refer to *The AI Homeschool Revolution* or contact us through *Secret Life of Learning* on Substack.

About the Authors

***Cassandra Jens**, a former Silicon Valley engineer turned home educator, helps families harness AI for personalized learning. She authors the **Empowered Homeschool Series** and writes about education innovation at **Cassandra Connects**.*

***Whitney Whealdon** is Director of Learning Innovation at Learning Tapestry and co-founder of Wonderwood. She writes **The Secret Life of Learning** Substack, exploring knowledge, reading comprehension, and AI. Whitney blends systems thinking, curriculum design, and literacy expertise to help educators and families build meaningful, knowledge-rich learning experiences.*

Chapter 12

Flourishing Through Change

Leading Higher Education with Purpose in the Age of AI

By Jennifer W. Shewmaker, PhD

Introduction: A Transformational Threshold

WE ARE LIVING through a moment that is reshaping how we teach, how we work, and how we understand essential human skills. Generative AI is not the first technological shift to reach higher education, but it may be the first to arrive with both the speed and the cultural impact that forces us to reconsider the nature of human contribution itself. For many educators and leaders, this has created a kind of existential tension. We feel responsible for preparing students for a world that is rapidly changing, yet we also feel the weight of traditions and practices that have shaped our disciplines for decades.

As the provost of a mission-centered liberal arts institution, I have engaged this tension in real time. Faculty members have shared stories of worry and loss. Staff members have wondered how AI will change their roles. Students have arrived with questions about authorship, learning, and responsibility. Across the institution, some have been excited to experiment; others have felt disoriented; and many have been simply unsure what to think.

It is vital to remember that, in these moments, fear is not a failure. It is a signal that something important is happening. When approached with care, fear can even become an invitation. It encourages us to slow down, to reflect, and to design intentionally.

This chapter grows out of the work that the team at Lipscomb University has been doing for the past two years. It reflects the model we developed to navigate AI with clarity and integrity, incorporating the *Fear to Flourishing* pathway, our institutional model for AI integration within the higher education context. This chapter also reflects a broader approach to hospitality-centered leadership and how that approach has impacted the process of AI integration. Most of all, it reflects a commitment to developing environments in higher education where people can flourish even in times of uncertainty.

The goal of this chapter is to offer a clear, hopeful, and grounded way to lead communities in higher education through technological change. AI may accelerate processes, produce ideas, and generate content, but it cannot replace the need for human wisdom, compassion, and purpose. These remain central to leadership and the process of learning, and when we design systems that honor these qualities, our institutions become more capable of navigating whatever changes come next.

The Landscape of Change: Why AI Feels Different

The rapid rise of generative AI has exposed a reality that leaders in higher education often understand intuitively. Change is not only technical, even when it is focused on emerging technology. It is emotional, relational, and cultural. The tools that are developing may be new, but the experience is deeply human.

When generative AI entered public conversation, many faculty and staff at Lipscomb expressed early excitement. They saw possibilities for creativity, efficiency, and new forms of student engagement. Yet beneath that curiosity was another set of emotions: confusion about what AI can and cannot do, grief about potential changes to long practiced teaching methods, and a sense of pressure to learn a tool that felt simultaneously powerful and unfinished.

This complexity often went unnamed in public discourse, which sometimes framed reactions in binary terms: enthusiastic early adopters on one side and resistant skeptics on the other. But as a psychology professor, it is clear that this framing oversimplifies what faculty and staff were experiencing. The truth is that nearly everyone felt both curiosity and worry. Many also felt alone in these feelings.

Naming emotional reality is a critical first step in helping communities move forward. We must understand the context that shapes adults' responses to disruptive technologies, including faculty, staff, and students. What we've come to understand, as we at Lipscomb have walked through this journey with AI, is that people rarely resist change for the reasons we assume. More often, they resist because their experience has not been acknowledged, their questions have not been answered, or they fear losing what matters most to them.

AI highlighted a profound series of questions, including: What is the heart of teaching? How do we evaluate learning? What does human creativity mean? How do we prepare students for a world where tools can generate content in seconds? The magnitude of these questions requires a leadership approach that is steady, relational, and rooted in mission.

Hospitality-Centered Leadership Systems

Mission-oriented formation is the practice of educating students not just for competence, but for purpose. It recognizes that learning is not merely the acquisition of skills or credentials, but the development of a whole person, intellectually, ethically, and spiritually, in alignment with the institution's deepest commitments. This means asking not only what students are learning, but who they are becoming in the process. Formation-centered institutions believe that education shapes identity, values, and vocation, not just résumés. Hospitality sits at the heart of institutions that focus on mission-oriented formation, and it is also a powerful leadership principle.

Effective teaching in higher education incorporates pedagogical strategies that create hospitable learning environments (National Research Council, 2009). The foundation of such environments begins with seeing the classroom as communal. As hooks (1994) observes, *"Seeing the classroom as a communal place enhances the likelihood of collective effort in creating and sustaining a learning community"* (p. 8). Within a communal classroom, everyone contributes to learning: not as isolated individuals performing for evaluation, but as participants in shared meaning-making.

Yet hospitality requires more than simply inviting participation. hooks (1994) proposes what she calls *"ruptures in the established order"* of higher education classrooms, so that *"everyone's voice can be heard, their presence recognized and valued"* (p. 185). This *"coming to voice"* recognizes the unique lived experiences, values, and beliefs each person brings and provides space in the classroom for all to be heard with respect and full recognition as part of the community. Palmer (1998/2007) describes this process as *"listen(ing) to those voices even before they have spoken"* (p. 47): making space for the other, being aware of the other, paying attention to the other, and honoring the other.

When we call students into shared responsibility for learning and create classroom environments that respect and provide opportunity for each voice to be heard, we increase students' ability to feel safe and take risks. As Palmer (1998/2007) puts it, *"If we regard truth as emerging from a complex process of mutual inquiry, the classroom will look like a resourceful and interdependent community"* (p. 52).

In the classroom, hospitality invites us to create learning environments where students can show up fully, bring their questions and personal strengths, and know that they belong. In leadership, hospitality invites us to design systems that give people space to learn and grow while honoring their dignity while using these same practices of inviting participation, giving voice, making space, paying attention and honoring the other.

A hospitality centered approach reflects four core commitments.

1. A Commitment to Presence

People need accessible leaders during moments of change. They need communication that is steady, transparent, and relational. Presence communicates care. It also helps leaders understand what their people are truly experiencing.

2. A Commitment to Agency

Hospitality honors the voice and expertise of the community. Faculty and staff want to contribute to decisions. They want to understand how changes will affect their work. Students want to be part of the conversation about the future. Agency strengthens both ownership and trust.

3. A Commitment to Transparency

Ambiguity fuels fear, and transparency reduces it. Leaders must provide consistent information about what is changing, why it matters, and how it aligns with mission.

4. A Commitment to Values

Mission-rooted institutions cannot adopt technology uncritically. Remember, all institutions have mission statements, so review the one at your institution as you consider this approach. Hospitality-centered leadership guides decisions by asking: Does this innovation support learning, human dignity, equity, and flourishing? How does it strengthen our purpose? How does it protect what is essential in regards to our mission and identity?

It's important to remember that hospitality is not passive, but rather requires intentional design. We understand hospitality as the practice of creating structures that support flourishing. In the age of AI, hospitality-centered leadership systems guide institutions to respond to change with clarity rather than panic, and with purpose rather than pressure.

We must be clear-eyed: when the pace of change is so quick, these commitments are difficult to make and to stand by. We have struggled with them ourselves, which is why we're recommitting to their importance as we share them here.

As institutions consider these commitments, it helps to think of them in terms of institutional mission and purpose. How might a hospitality-centered approach be adapted for this specific institution? Take a look at the mission statement, core values and vision statement for the institution and consider how each of these commitments might fit best in that context. Doing this work to consider the fit for the individual context will allow this approach to best serve the community. Hospitality-centered leadership systems become the scaffolding through which an institution can move toward flourishing, even in times when the future feels uncertain.

Hospitality in Practice: Designing for Human Flourishing

As we implemented the *Fear to Flourishing* model I'll share below, three practices that flow from the commitments consistently strengthened our community's capacity to flourish.

Practice 1: Naming Reality with Compassion

People need clear and honest communication. They also need reassurance that they will not face change alone. When we named both the challenges and the opportunities of AI, we created space for people to process their experiences. This practice built trust and made adaptation feel possible. We've found that this is a continual process that must be approached with all stakeholders. This practice flows out of our commitments to presence and transparency.

Practice 2: Fostering Agency and Voice

We encouraged faculty and staff to shape the direction of our work. Their expertise was essential and their questions guided our training. Their experiments informed our policies. Agency helped people feel empowered rather than threatened. This practice is connected to our commitment to agency.

Practice 3: Building Predictable Structures for Trust

Trust grows when communication is consistent, expectations are clear, and values are visible. We supported these structures through regular updates, ongoing listening sessions, clear guidelines, and a culture of learning. This practice flows from all four of the commitments above, as we provide presence, transparency, agency and a consistent focus on the values of the institution.

These three practices have helped us design an environment where people could develop skill, confidence, and resilience. It is still a work in progress, but we are committed to continuing these practices to lead to flourishing for all.

Understanding the Human Landscape: The Fear to Flourishing Pathway

Shortly after the widespread launch of generative AI tools, our leadership team needed a way to talk about the feelings and questions that were emerging across campus. An AI Task Force was formed in January 2024, and we shared a report in July 2024 with recommendations for next steps. Sarah Gibson, Director of the School of Communication at Lipscomb, served on the AI Task Force. She was then appointed as the AI Faculty Fellow in the Center for Teaching and Learning to aid us in implementing the Task Force recommendations. As a part of this work, Gibson created the *Fear to Flourishing* pathway to explain how people move from fear to flourishing (in preparation). This framework has been helpful in providing a framework that connects emotional experience with professional growth.

The *Fear to Flourishing* pathway outlines seven stages that individuals typically move through in times of technological transition. These stages are not linear and people move through them at different paces. Each one offers insight into the emotional work that must accompany the technical learning that comes with integrating AI into our curricular and technological systems.

When we see people resisting change, we've come to understand that there is usually a missing component in the change environment as described by Knoster (1991) in the model for managing complex change. Knoster posits that successful change requires five elements: vision, skills, incentives or motivation, resources, and a plan for action. When any of these elements are missing, people tend to be resistant. Depending on where a person is in the individual roadmap above, understanding what may be missing helps us in supporting individuals to move through the change process. For example, if someone is missing a skill, we can help them build it. If a faculty or staff member does not have an action plan, we can support them in developing one.

Stage 1: Initial Understanding

At this stage, people are aware that something significant has changed but have little clarity about what it means. Questions often sound like: What is ChatGPT? Does this replace us? How much do I need to know right now?

Leadership Purpose: Provide simple, accessible explanations that give people a foundation to begin to understand what AI is and provide context for this technological change along with others that have impacted education (for example: the printing press, the industrial revolution, the rise of the internet, mobile technology).

Stage 2: Grief

Grief is an honest and healthy response to change. Faculty who have refined their craft over many years may feel that their expertise is being overshadowed or undervalued. Assignments that once assessed student learning effectively may need to be reimagined.

Leadership Purpose: Make room for grief. Without that space, people may become stuck in fear. Acknowledge that some may be feeling a sense of loss while others may be excited for the possibilities that are emerging, and all of those feelings are acceptable.

Stage 3: Day of Reckoning

This stage arises when people realize that their assignments, policies, or workflows no longer function as intended. This can be overwhelming, and it can also be transformative as we face the inevitable challenge and begin to imagine what comes next.

Leadership Purpose: Offer tools and support to help people reimagine their work with integrity.

Stage 4: Flip the Script

Once the initial shock has settled, curiosity begins to emerge. People start asking new questions, such as: Could AI help me explain a concept more clearly? Could it support students who struggle with writing? Could it streamline processes so I can focus more on mentoring?

Leadership Purpose: Nurture these questions by offering low-pressure opportunities to experiment. At Lipscomb we created faculty sessions in the Center for Teaching and Learning where people came to share new ideas, challenges, and solutions with one another.

Stage 5: Curiosity

At this stage, people begin to play with AI. They try prompts, test tools, and share small successes. Learning becomes more collaborative. This stage marks an important shift from fear to possibility.

Leadership Purpose: Provide space for conversation. The sessions in our Center for Teaching and Learning grew beyond idea sharing to allow faculty to try new ideas and report back, giving each other feedback and sharing results with one another.

Stage 6: Integration and Sharing

Here people begin incorporating AI into their regular work. Faculty share assignments and strategies. Staff streamline tasks.

Leadership Purpose: Look for a rise in collaboration and excitement. It's important to consider ways to support this by providing a platform to celebrate things that are happening on campus to inspire others to reimagine.

Stage 7: Honing

Eventually people develop more disciplined and ethical AI practice. They can distinguish when AI is useful and when human expertise must remain central. They refine their approach based on mission, values, and evidence.

Leadership Purpose: Focus on providing ways to share refinements and spotlight mission and values linked to the outcomes of this work.

Understanding this pathway helped us design leadership strategies that supported people at each stage. It reminded us that emotional work and technical work must move together. And, It also reinforced the importance of hospitality, because people cannot flourish if they do not feel seen or supported.

Building Institutional Capacity: Lipscomb's Seven Step Model

Our institutional model for AI integration grew from both research on change management and our lived experience with faculty and staff. The seven steps below describe a coordinated approach that centers people while building sustainable structures.

As you consider the model we adopted, think about your own institution and how it might be adapted for its context. For example, at Lipscomb, the Provost led the AI Task Force, while at a larger institution an Assistant Provost may lead it, or perhaps a faculty member who is well respected and knowledgeable about technology and the institution. The key is to use this model as a starting point to consider institutional context and what would work best for it.

Step 1: Launch a Representative AI Task Force

As we were creating the AI Task Force, our President knew that it needed broad representation. Faculty from multiple disciplines, technology leaders, ethicists, instructional designers, student representatives, and administrators all joined the initial team. Their charge was to learn, test tools, ask critical questions, and design resources and recommendations that would be the foundation of our integration. This group provided both expertise and relational trust.

Our first task was to identify our own guidelines in relation to AI. Lipscomb has eight core values that guide our work, so the team used those to create our unique approach to AI integration. These core values are: Love God, Serve Others, Respect All, Pursue Joy, Seek to Learn, Embrace Collaboration, Deliver Our Best, and Create Solutions (Lipscomb Strategic Vision, 2022).

When considering a specific institution, think about the mission statement, core values and vision that have been articulated and consider how AI use relates to those foundational pieces of institutional identity.

Step 2: Host a Visible AI Event

We learned early that training works better when people feel connected to the purpose behind it. We used our annual opening faculty meeting in August of 2024 to begin talking about AI and its integration with the strengths and values that we hold. This allowed faculty to see demonstrations, attend workshops, and hear from experts and colleagues. This event helped shift the institutional narrative from anxiety to possibility. We followed this up with monthly trainings, faculty interest groups, a regular AI newsletter, a second all-faculty meeting in August 2025, and a focus on integrating AI with the timeless wisdom of the liberal arts in each monthly faculty meeting.

When considering a specific institution, identify several key regular events that might be used as starting points for this conversation. Is there a meeting at the beginning or end of each semester or quarter? Does the Center for Teaching and Learning at the institution offer regular training sessions? Is there an online resource center that people make good use of? The idea is to build the AI conversation into the normal flow of the year so that people see how it fits within the institutional system.

Step 3: Develop Academic Integrity Guidelines

Policies must be clear, practical, and rooted in mission. Ours focused on transparency, responsible use, and respect for human dignity. We encouraged faculty to use AI in ways that support learning without undermining critical thinking or integrity. We also provided guidelines for how to tell students what was expected in terms of AI use for each assignment, providing transparency and teachable moments throughout the semester.

This is a critical step to provide clarity for both faculty and students regarding AI use. Work together as a community to decide what is acceptable and how it is going to be taught to the students.

Step 4: Offer Just in Time Training

People learn best when the training responds to a current need. We created short workshops, department level sessions, and one-to-one consultations. These sessions helped faculty redesign assignments, explore prompt strategies, and identify effective use cases.

Consider how the institution might best do this. For example, is there a central newsletter or website that could host some just-in-time learning information? Are there regular session opportunities that might be used to provide trainings?

Step 5: Support Growth and Spread

Once early adopters began to develop effective practices, we created ways for them to share what they were learning. Faculty showcases, panel discussions, and cross-campus presentations created momentum. Stories of success were powerful motivators.

How do faculty at an institution share stories with one another? Build on community practices to provide opportunities for faculty, staff, and students to share information, experiences, and successes with one another.

Step 6: Move Toward a Universal AI Platform

As usage increased, we realized the need for a university-wide system that met our standards for privacy, safety, and educational integrity. Selecting such a platform required careful evaluation informed by our values. As a university, we wanted to have the opportunity to share a variety of tools with our students, so that was something we were looking for as we evaluated different platforms. We also considered environmental impact and security.

For the institution under consideration, identify what is most important for that context, lay out those criteria clearly with the group who is making the decision on the tool, and then evaluate the candidate tools based on the specific needs within your context.

We chose BoodleBox, which has provided us with a single platform we can use to teach students how to work with several different AI tools; is open about their environmental impact and the work they are doing to minimize it; and creates a secure environment for exploration of AI.

Step 7: Re-Root in Mission

Throughout the process, we continued to return to our identity as a mission-centered institution. Our values guided decisions about what AI should and should not do. For example, in our Center for Teaching and Learning, we have provided faculty with the opportunity to consider how and if they should use AI in each course, and then supported them in creating explicit instructions to help students understand the expectations for AI use within different assignments. Faculty were coached on how the syllabus should include at least the following considerations: under what circumstances AI use is permitted or forbidden; if AI is permitted, how students should cite, credit, and document use of AI tools; and guidance on how to use AI as a tool to learn. Materials should include a warning about the technology's limitations and clear rules regarding students' accountability for AI output, and AI tools should be used ethically and responsibly, along with a discussion of what that means in the course.

The institution's core values also shaped our commitment to protecting human dignity, creativity, and agency. Every institution has a mission, so the key is identifying how AI integration connects with yours.

These steps provided a structure that supported emotional, academic, and organizational change. They created a clear path forward without rushing the process or leaving people behind. Now, to be realistic, this is not easy, and we still have small groups who are struggling, be they faculty, staff, or students. However, the key is to keep the *Fear to Flourishing* model in mind, value transparency, and continue to give people room to grow toward flourishing.

In considering a specific institution, think about what steps need to be put into place to integrate AI. Do they look like these? Are there other steps that might be important in that context? The questions above should help leaders in higher education think through each step within the institution, understanding that change is contextual, and doing the work of considering the next best steps is vital for each organization.

Below are the outcomes that we've experienced as we've gone through this process.

Institutional Outcomes: What Flourishing Looks Like

As our approach matured, several outcomes began to emerge across campus through faculty, staff, and student surveys, monitoring of AI usage via Boodlebox, and focus group sessions.

Deeper Faculty Confidence

Faculty who were once skeptical now report that AI supports their teaching, expands their creativity, and strengthens student engagement. They are sharing new ideas with one another and using new tools, demonstrating agency in using AI in their classrooms.

More Ethical and Mission Aligned Practice

Rather than using AI uncritically, faculty are developing thoughtful approaches that prioritize formation, integrity, and human presence. We have put discussion groups in place in the Center for Teaching and Learning that allow for deep discussion of ethics and have developed shared resources to guide student discussion of ethics, grounding all of our work in our mission.

Greater Academic Innovation

Departments are designing more robust assignments that emphasize higher order thinking, synthesis, collaboration, and discernment. Through an AI and pedagogy faculty discussion group that we created in the Center for Teaching and Learning, faculty members are able to share new ideas, try new tools, and support one another in engaging with new technology.

Stronger Community and Collaboration

AI has become a shared area of curiosity. Faculty, staff, and students collaborate more readily and learn from one another. Our university is very relational, and faculty have been able to use AI to work in community with their students as they use new tools and create communities of learning amongst the faculty.

Sharpened Institutional Mission

AI has not weakened our mission or identity. Instead, with intentionality, it has highlighted our commitment to human dignity, hospitality, and formation. As a community, we talk about AI use often, and provide opportunities for faculty, staff, and students to learn from one another. We put the mission front and center through linking our AI use with our core values.

These outcomes emerged not because AI is extraordinary, but because our leadership approached AI with purpose, clarity, and care.

Conclusion: Leading with Purpose in a Time of Acceleration

AI in higher education is standing at a threshold. The choices we make now will influence how future students learn, how faculty grow, how staff function, and how our communities define human flourishing. **AI is powerful, but we have come to realize that it is not the center of this story: people are.**

When institutions lead with hospitality, clarity, and courage, technological change becomes an opportunity for growth rather than a threat. When we ground our decisions in mission, and design systems that honor the worth of every person, we create environments where people can flourish even in the face of uncertainty.

Flourishing in the age of AI is possible. It begins with seeing people clearly, designing with purpose, and leading with a steady hand. It is not the absence of fear, but the presence of hope. And, it is not driven by urgency, but by conviction.

AI may change how we work, but it does not change what matters most. Our calling as educators is to nurture wisdom, curiosity, compassion, and character. When we protect and strengthen these qualities, our institutions will not simply survive change; they will flourish through it.

***Acknowledgement**: The Fear to Flourishing framework, its associated concepts, and roadmap language are attributed to Sarah Gibson and are used with acknowledgement in this chapter with citation of an upcoming publication.*

About the Author

__Jennifer W. Shewmaker, PhD__ is Provost at Lipscomb University in the United States, where she leads strategic initiatives around faculty development, AI integration, and liberal arts curriculum reform. Her work focuses on hospitality-centered leadership and designing systems where people can flourish in times of disruption. She has written multiple book chapters and articles on teaching and learning and shares weekly thoughts through a Substack newsletter __jenshew.substack.com__. Dr. Shewmaker speaks nationally on higher education leadership, student formation, and institutional transformation in the age of AI.

Chapter 13

Why Women Must Get AI Savvy Now

By Cheyenne Dominguez

I REMEMBER WHEN I first tried ChatGPT in mid-2023. I was sitting at my kitchen counter trying to write an important email, and even though I enjoy writing, for some reason the words just weren't landing the way I wanted. I opened ChatGPT to help me, thinking it might give me a little nudge. It wasn't perfect, but it definitely got me unstuck, and I remember pausing and thinking, *"Wow, a shift in technology is happening."* One minute you're just asking for a quick draft so you can move on with your day, and the next you realize these tools are becoming part of how we work, think, and make decisions.

Since then, in almost every conversation I've had with women through my work as the founder of M(AI)VENS, the AI newsletter and network for non-techy women, someone will say they know they should start learning AI, but feel hesitant or worried they'll somehow get it wrong. I hear that a lot, and I get it. As women, we carry so many responsibilities at once that adding one more thing, especially something that feels technical or unfamiliar, can feel exhausting.

Historically, women have stepped into new technology later than men, and there are real structural and cultural reasons behind that pattern.

Across decades of digital innovation, women have often been positioned as later or less frequent adopters of new technologies than men, especially in the early stages. A large review of IT adoption studies from 2000–2017 found that men were more likely than women to try new technologies, including shopping online, and reported higher confidence and more positive evaluations of websites (Shaouf, A. & Altaqqi, O. 2018).

Ethnographic work in public internet access sites similarly showed men comprising about 60% of users versus 40% women over a ten-year period in one U.S. city, revealing how "tech spaces" have long been subtly gendered (Dixon et al., 2014).

These gaps reflect historical patterns in lack of access, time, confidence, and social norms that made it easier for men to experiment with each new wave of technology first. When women arrive later, they enter systems and markets already shaped without their perspectives, limiting both opportunity and influence over how tools are built and governed. In the context of artificial intelligence now being woven into hiring, finance, healthcare, and creative work, repeating this pattern would mean letting others design the future and then being asked to adapt to it, rather than helping to define it from the start.

What concerns me is that many women don't realize how much they stand to lose by waiting. It's not just about staying current. Our understanding of AI will affect income, visibility, confidence, and the ability to lead. I see women who are brilliant at what they do, yet they're being quietly outpaced by competitors or colleagues who aren't necessarily more talented, just more willing to experiment with these tools.

But I also know this. Many women bring a careful, thoughtful, steady approach to new things. We ask more questions. We consider impact. We're great collaborators. We think about fairness and long-term outcomes. Those instincts aren't obstacles in AI. They're strengths that help us use these tools with intention instead of getting swept up in hype.

My hope is that this chapter gives you permission to start exactly where you are and feel supported while you do it.

The Cost of Slower Adoption

Recently, it's been nearly impossible to ignore how quickly AI is reshaping the way people work. It's showing up in job postings, performance expectations, marketing strategies, leadership conversations, and almost every corner of professional life. You can feel the acceleration happening in so many industries.

When I talk with women about this shift, many tell me they know AI is important, but they feel unsure about how to begin. Some hope they can wait until things settle, or until they feel more ready, or until their organization offers training. The challenge is that nothing is slowing down. While some women are waiting for the right moment, others are experimenting, learning, and building confidence.

One of the clearest places where slower adoption carries a cost is income. AI is helping people streamline tasks that used to take hours, whether it's drafting proposals, preparing presentations, writing marketing content, or organizing client information. When we as women don't feel ready to try AI tools, we may find ourselves doing everything manually, which limits how much we produce in the same amount of time. Over the long run, that gap can influence earnings, opportunities, and even how we as women price our work. Meanwhile, those who use AI well can take on more projects, respond to clients sooner, and deliver work that feels polished and thoughtful without sacrificing their nights and weekends.

Slower adoption also affects visibility, which plays a bigger role in women's careers than we tend to talk about. It takes consistency to stay top of mind, whether you're building a business, growing a personal brand, developing as a leader, or seeking new opportunities. That means sharing ideas, contributing to conversations, and showing up regularly.

Confidence is another area of opportunity when it comes to AI. Women who experiment with AI early tend to feel more comfortable speaking about it. We show up in meetings with a sense of readiness because we've seen how the tools work in real time and know we can participate in the conversation. Women who wait often tell me they feel a little out of step, like everyone else is speaking a language they haven't learned yet. Of course, that feeling has nothing to do with ability. It's simply a result of not having enough exposure. Over time, that limited exposure can negatively influence how often women speak up, take risks, or share ideas that deserve to be heard.

Leadership is another place where slower adoption has a ripple effect. AI is influencing high-level decisions in every industry, whether it's about productivity, innovation, communication, or ethical use.

Those who understand AI, even in a basic, practical way, often become the ones who help shape those decisions. Women who get comfortable early become valuable contributors to conversations about how their organizations should approach AI. They help guide the choices, ask thoughtful questions, and advocate for policies that reflect the needs of real people.

Harvard Business School researchers argue that if women remain underrepresented among AI users and contributors, three main harms follow:

- women fall behind in AI-complementary skills and career progression;
- models trained mostly on men's inputs risk encoding male-centric assumptions; and
- organizations miss productivity and innovation gains that come from broader participation (Blanding, 2025).

Another consequence of slower adoption is how it affects professional reputation. When AI is becoming standard in how people prepare and present their work, women who aren't using it may unintentionally look less knowledgeable or less prepared, even when just as capable as their peers.

One approach is to treat AI as a private thought and preparation tool rather than a shortcut. Women can use it to organize complex information before a meeting, pressure-test an argument, or refine language to feel more persuasive. Consider leveraging AI for output that reflects stronger thinking and clearer communication. Over time, this helps women benefit from AI technology while maintaining ownership over their work.

There is also a cost in terms of career mobility. More job descriptions mention AI every day, even for roles that have nothing to do with engineering or data science. Employers want people who can use AI as a partner in their work. Women who start learning now have the opportunity to demonstrate how they're proactive, adaptable, and ready for the future. Women who wait might find those roles they once felt qualified for now expect familiarity with AI tools they haven't had a chance to explore.

And perhaps one of the most subtle, but meaningful, costs is influence. Influence comes from having a voice in shaping how new systems, tools, and decisions unfold. AI is becoming part of the foundation of modern business, and the people who understand it are the ones shaping that foundation. When women hesitate, their input is missing in moments where it could make a real difference. That's how systems drift away from the values women often champion, like fairness, accessibility, and long-term wellbeing. When women engage early, even at a basic level, we can help ensure that AI is used thoughtfully, and that decisions reflect the needs of more than one type of person.

The good news is that none of these costs are permanent or inevitable. They only exist when women step back instead of stepping in. So, let's close the gap.

The cost of slower adoption is real, but it's not unavoidable. Starting now gives women the advantage of learning while the world is still figuring this out, which is exactly when new voices are needed most.

The Unique Qualities That Position Women to Lead in AI

One thing I've noticed in all my conversations with women about AI is that many of us underestimate how perfectly designed we are for it. There is a common belief that AI belongs to people who are super technical, more analytical, or more comfortable with machines than with people. Yet when I look at the traits that make someone successful with AI, I see qualities women use every single day. They're everyday ones that come from years of navigating complex situations, solving problems, and carrying responsibilities that require judgment and awareness.

Women tend to move through decisions with care and thoughtfulness, especially when the stakes involve people or long-term outcomes. That approach is incredibly valuable when the world introduces something that has the potential to reshape so many areas of life and work. I'm not saying women are too cautious or risk-averse, but rather that women often think about the ripple effects of their choices. They want to understand how something will impact the people they lead, the people they serve, and the people they love. That instinct translates beautifully into AI, because the people who use it well aren't the ones who jump in blindly. They're the ones who pause, are thoughtful in their decisions, and ask the right questions.

I've seen this instinct show up in boardrooms and nonprofit meetings, in volunteer work and executive roles, in entrepreneurship and motherhood. Women often hold the big picture in mind while managing the small details. That ability to see the whole landscape while staying connected to the day-to-day is exactly what AI requires. These tools can be incredibly helpful, but only when paired with someone who pays attention to context, values, and impact. Women naturally bring that perspective, even if they don't always recognize it as a strength in leadership.

Another quality women bring into this chapter of innovation is collaboration. Women tend to share what they learn, support others as they figure something out, and build connections around new experiences. AI grows stronger when people talk to each other about what works and what doesn't. Women have been practicing that kind of community learning their entire lives, often in ways that go unnoticed because it feels like second nature.

Women also have a strong ability to read nuance. We sense when something feels off, right? We notice when a message doesn't land right. We understand when a decision might create unintended consequences. AI isn't perfect, and it certainly doesn't replace intuition. That kind of instinct builds trust, and trust is the currency of leadership in any field.

There's also a level of resilience that women carry, often quietly. Many women know what it feels like to figure things out on their own, to learn by necessity, or to build skills while balancing more than one role. That resilience translates directly into AI learning. You don't have to be perfect. You don't have to understand everything at once. You only need the willingness to start and the patience to keep learning.

Another strength that matters right now is women's ability to think about relationships, communication, and community. AI is transforming how we connect with people—through writing, messaging, marketing, and storytelling. Women often lead in these areas because they understand human behavior. We understand what people need to hear, how they want to feel, and what builds connection. AI is a tool, not a voice. It's the human behind it who shapes the message. Women are strong storytellers, strong communicators, and strong bridge-builders.

Women often lead with a blend of practicality and care that considers more than efficiency or productivity. We think about fairness. We think about safety. We think about whether something aligns with our values. These qualities matter deeply in an AI-driven world, where decisions move quickly and the consequences are still emerging.

When women recognize these strengths as assets, everything shifts. AI stops feeling like a technical mountain and starts feeling like a tool that fits naturally into the way we already think and lead.

I believe there is a confidence that comes from realizing you aren't starting from zero. Women bring life experience, emotional intelligence, discernment, creativity, leadership, and endurance into everything they do. These intrinsic strengths are the very traits that can help shape how AI is woven into our workplaces, families, and communities. Women don't need to become techy to participate in the AI revolution.

The Opportunity in Front of Women

There is something so interesting and meaningful about where we are in this moment of technological change. AI is growing quickly, but the landscape hasn't settled. Roles and expectations haven't hardened. People are still learning, experimenting, and comparing notes. Women have room to explore without feeling behind, and that kind of window doesn't appear often.

There's an unusual level of access available right now, particularly access to free or low-cost AI tools, open learning sources, and public-facing platforms to explore AI without a large investment. Cost and time are real considerations, especially for women balancing full lives, and they help explain why adoption has sometimes lagged.

Because many organizations are still designing norms and practices, women have an opportunity to influence how their workplaces approach AI. When women understand enough to participate meaningfully in those discussions, they help bring balance and a wider lens to decisions that might otherwise be shaped from a narrow point of view.

Many women also carry ideas they've tucked away for years. A book they want to write. A business they want to build. A program they want to design. Often, the early stages feel too heavy to take on. Women can look for ways AI can lighten that first lift. It could help shape the structure, outline the concept, or turn scattered thoughts into something more tangible. When the beginning of a project feels less overwhelming, women are more able to bring their ideas to life.

What stands out most to me is the amount of room women have to shape their own path with AI. We can choose the pace, the tools, and the areas that matter most to us. We can learn, test, explore, and adjust without the pressure of being perfect or ahead of anyone else. Our strengths, thoughtfulness, clarity, collaboration, and resilience, all fit so naturally into this landscape.

If you're reading this and feeling like you're starting from zero, with no technical background and no idea where to begin, the easiest entry point is simply exposure.

Start by subscribing to a few AI newsletters that relate to your industry so you can begin to recognize key players, recurring tools, and the kinds of problems AI is being used to solve in your line of work.

Follow AI voices and platforms on Instagram or LinkedIn, including accounts like ChatGPT, Claude, Perplexity, and Gemini, which often share product updates and new features in plain language before they show up anywhere else.

Pay attention to AI investment trends as well. Where significant money is flowing, advancements and consumer-facing products usually follow.

From there, experiment. Try ChatGPT, Claude, or Perplexity using the free versions first, just enough to get a feel for how they think and respond and where they might be useful in your day-to-day work. If you reside outside the U.S., you may encounter Mistral or DeepSeek, which are gaining market traction in parts of Europe and Asia. You don't need to try all at once. Pay attention to how these tools think, how they handle nuance, and where they fall short. When one starts to feel genuinely useful, consider upgrading a single tool to the paid version and notice how much more capable it becomes.

You can also take advantage of free learning options through Google AI, IBM, or LinkedIn Learning, which are designed for non-technical users. It's about building familiarity, confidence, and a sense of orientation so AI starts to feel like something you're engaging with, not something happening around you.

The opportunity in front of women right now is not tied to mastery or expertise. It comes from stepping into the moment while it's still flexible, while the entire world is still learning, and while women's voices are needed to help guide the direction ahead.

A Note From Me to You

As you finish this chapter, I hope it leaves you feeling excited and encouraged. AI can feel overwhelming from a distance, yet it becomes much easier to approach once you start leaning in.

I still think about the morning I first opened ChatGPT at my kitchen counter. I wasn't trying to move mountains. I was just stuck on the wording of an email. That experience led me to create M(AI)VENS, the fast-growing AI newsletter and network for non-techy women, to learn together about what this technology means for our careers and our daily lives.

Begin from exactly where you are, using the strengths you already depend on in every other part of your world.

My hope is that you carry forward a sense of possibility. You are more prepared for this AI revolution than you might realize, and I'm cheering for you as you take your next step.

About the Author

Cheyenne Dominguez** is the founder of **M(AI)VENS**, the AI newsletter and network for non-techy women. Cheyenne has built programs that empower women to lead, connect, and grow. Recognized as a Power Woman in Greater Philadelphia and one of Miami's 50 Savviest, her work through M(AI)VENS was featured in **The Hustle**, reaching over 2 million readers. She writes and speaks on women's leadership and is passionate about helping women stay competitive, visible, and influential. Visit **www.maivenscommunity.com

BUSINESS

Chapter 14

AI For Solopreneurs

Rethinking What Help Looks Like

By Anonymous

YOU'VE BEEN STARING at your screen for forty minutes. The decision should be simple. You've gone through this process a million times before: choose one vendor over another, approve or delay a launch, respond to that difficult client email. But for some reason, your brain is sludge. Your gut is no help. Maybe the client who cancelled earlier threw you off. Maybe it's the family fallout from you missing your daughter's dance recital. Maybe it's the stress of wondering if you'll make payroll next week.

For a million great reasons, you're too tired to think clearly, and too far behind to push it aside. You're supposed to be the visionary, the strategist, the confident one who knows what they're doing.

Instead, you feel alone. Frustrated. Trapped in a decision you don't feel equipped to make.

This is what solo entrepreneurship (being a "solopreneur") really looks like, more often than anyone wants to admit. Entrepreneurship takes many forms, from individuals running their own single operator business (such as consulting, and agency, or a service like bookkeeping or computer repair), to larger companies that employ multiple employees. Running your own show as a solopreneur can sound liberating to others, but behind the polished pitches and curated success stories is often a daily grind of uncertainty. Every decision seems to involve so many variables, yet you never seem to have enough data, and often there's nobody to talk to who *really* gets what you're up against, when you're up against it.

This is where AI can help. Not because it's fast, or cheap, or convenient, but because it can meet you exactly where you are. Without judgment, lectures, or that annoying *"I told you so."*

We're all used to hearing about AI in terms of automation… streamlining your operations, generating content, or analyzing market trends. And yes, those are useful features. But if you're a solopreneur, you already know that the hardest part of your job isn't just doing the work.

It's figuring out what work to do in the first place.

It's deciding. Prioritizing. Planning. Course-correcting. Weighing pros and cons… not to mention figuring out which people are professionals and which are con artists. It's thinking through the consequences of a thousand small choices, each of which could shift your trajectory by 2% or 20%.

And when you're tired, uncertain, or stuck, that's when bad decisions get made. Or worse, no decision gets made at all.

Now imagine having a calm, clear, always-available thinking partner who doesn't get tired, doesn't get emotionally reactive, and doesn't judge you. One that can test your thinking, reflect your logic back to you, challenge your assumptions, and suggest alternatives, or simply sit with your uncertainty without rushing to fix it.

That's the deeper, more transformative role AI can play in your solopreneurial life: not as an outsourced assistant, but as a *cognitive partner* that helps you think better, see clearer, and act more decisively, even in the fog.

This chapter is about something more foundational than the latest AI tools or a roundup of trending apps. This chapter explores how AI can help you *do the hard parts* of running a business: thinking under pressure, making sense of complexity, managing emotional situations, and learning from your own patterns and the patterns of your business.

If you're going to thrive, not just survive, as a solopreneur, you need more than automation. You need support, resilience, and clarity. You need to stop thinking alone.

AI As Strategic Collaborator: Beyond Productivity

As generative AI has rolled out across the consumer and business world, we've been trained to think of it in task-oriented terms: writing, summarizing meeting notes, automating client outreach and booking. While those are all valuable, there's much more to it than that. AI isn't just a faster set of hands, an extra employee who works for free and never sleeps. It's also the thinking partner you didn't realize you needed.

This shift from tactical tool to strategic collaborator can change everything. You're not forced to muddle through quandaries with limited information all by yourself at 11:30 on a Thursday night. With AI, you have access to different points of view, contrarian perspectives, and an intelligence that can surface considerations and factors you might not think of, yourself, when they matter most. Instead of being locked into a limited decision-making process that could have real consequences for your business, you can get the help you need to not only surface more areas to consider, but also weigh the pros and cons for the short and long term.

Let's say you're considering a major pivot. You're weighing a new service line, a change in pricing, or how to better engage with customers. You've done some research, talked to a few trusted contacts, and now it's just you, sitting at your desk, trying to make sense of a mishmash of data and conflicting instincts.

This is where AI excels, not by delivering the "right" answer on the spot, but by helping you think in ways that lead to better answers. You can ask it to map out the upsides and downsides of each option, highlighting unintended consequences. You can explore how similar choices played out in other industries, surfacing blind spots or biases in your reasoning. You can even reframe your decision through different lenses (customer experience, operational load, brand positioning, etc.), maybe finding multiple viable options, rather than just one.

In short, AI helps you think *like a strategist,* even when you're running low on clarity and time. It lets you operate as an executive reviewing your options with a staff of AI "direct reports". It gives you the chance to review the available options, get input from your dedicated staff, and then decide on your best next steps, based on a much broader set of considerations than you'd have thinking everything through on your own.

In high-stakes, high-stress moments, that kind of structured reflection is often what separates impulsive, expensive decisions from wise ones that actually build your business.

No More Isolated Decision-Making in the Fog

Imagine you're choosing between two accounting software packages: both offer similar packages, both are priced within budget, and both come recommended. You're not sure whether to prioritize long-term scalability, ease of onboarding, or cultural alignment.

Instead of getting stuck in a loop thinking and re-thinking things from every possible angle (fondly known as *analysis paralysis)*, you can use AI to simulate different decision paths, exploring:

- What are the tradeoffs if you prioritize ease of onboarding over price?
- Which software can grow along with you to support your expansion for the next 12 months?
- A weighted decision matrix of features and functionality that lets you compare multiple packages according to your own criteria.

With plenty of context about your situation, relevant data, and an in-depth discussion, AI won't just give you an answer. It'll create a framework for *getting* to an answer. That's the cognitive support you didn't know you needed until you had it.

It's tempting to think of AI as a tool that *does* things. Yes, it does. Absolutely. But it also *thinks with you*. It lets you explore, without the pressure to be perfect. You can brainstorm, revise, rehearse, and revisit... all in the same space. You can test your messaging with customer personas, reality-check your logic, or even just practice saying something out loud before you bring it to a client, investor, or professional peer.

With strategic AI, you're not replacing your own human intelligence. You're extending it, creating more surface area for insight, more pathways for clarity, more momentum when you feel stuck.

Real Solopreneurs, Real Problems, Real Support

This isn't theory. Solopreneurs are already using AI in high-stakes, emotionally charged, financially critical situations. And they're seeing results that go far beyond time savings.

Now, AI isn't flawless and it's not necessarily smarter than a human. But it can provide a kind of mental scaffolding that holds a wide variety of pieces while you figure out the puzzle. And it does that without ego, pushing back just enough to help you hear yourself think.

Let's look at a few stories of individuals who were facing pressing business challenges alone who got the help they needed through collaboration with AI ... not through automation, but through strategic clarity, confidence, and concrete next steps.

Reclaiming 20 Hours A Week

A consultant working with an AI "Dream Team" didn't start out as a believer. Like many, she saw AI as a helper for tasks, not thinking. But after 30 years in corporate life, transitioning to independent consulting was wearing her down. With so much to learn, she had to get help.

So she built an AI "Dream Team" (a group of personas, each trained for different strategic roles) and everything changed. Her new team didn't just help her get more done. They helped her think better. With personas dedicated to research, brainstorming, strategic review, and tone-checking, she found herself delegating lines of thought, not just chores. The result? Walking into multiple client meetings each day with confidence. Crafting offers that made sense for her ideal customer. All this with more than 20 hours a week, which freed up more energy to focus on what mattered most.

AI didn't make her more efficient. It made her more present to her work, her insights, and her clients.

Saving The $5M Project

A multi-year, $5 million civic project was on the brink of collapse. One board member's resistance was derailing timelines, stalling approvals, and threatening the entire initiative. The lead consultant brought in a conflict-resolution AI Dream Team. They didn't just diagnose the politics, they unpacked the psychology. The resistant board member wasn't malicious; they were scared, overextended, and nobody felt comfortable confronting them.

With the AI's help, the consultant reframed the entire narrative, acknowledging past commitments while ushering in a renewed, collaborative vision. With the help of the Dream Team, the consultant generated a list of actionable steps that translated into real results. Tension eased. Consensus returned. Funding was preserved. The project not only survived, but also was strengthened by the crisis, all because someone had the right support at the moment they needed it most.

These use cases aren't edge cases. These are everyday entrepreneurial problems made solvable through better thinking, clearer reflection, and emotionally intelligent support. Without AI to assist, people could have solved those problems. But probably not as quickly. And not with the same kind of help *in the moment* that made it possible to understand the nuances of what was *really* going on.

Filtering The Noise – Choosing The Right Tools for You

If you're running your own business, chances are your inbox is filled with *must-have* software recommendations, your social feeds are flooded with AI tool reviews, and your friends are pinging you about the next big platform that promises to save you 10 hours a week. Everyone is selling something, and everything looks like a solution. But how do you know what will actually work for *your* business?

This is where AI strategy and function combine for real impact: *as a meta-tool.* Not just something you use, but something that helps you understand **what to use–and why**. This shift is critical. With so many options on the market (and more coming), many AI tools are *"solutions looking for a problem."* AI can help you figure out what problems you actually need to solve **for your business**, and then help you find the tools that let you do just that. Rather than chasing every shiny object, you can use AI to:

- Analyze your specific business needs and match them to available tools.

- Summarize and compare product features from multiple sources.
- Flag hidden costs or missing features that don't show up in top-level marketing.
- Simulate workflows to see how well a tool would integrate into your actual systems.
- Draft emails or pitch questions to vendors based on your real priorities.

In the end, with the right kind of focus, you may discover that your needs are far simpler than the tools being promoted. In fact, you may discover that AI isn't necessary for certain needs. Maybe you just need a process. Maybe you just need to clean up your data. Adding tools to a bad process just amplifies the issues and increases operating costs. A well-structured exploration with AI can prevent that all too common kind of pain.

The Tool You Didn't Know You Didn't Need

Imagine you're considering a scheduling platform. You've been planning to implement one, and everyone seems to love App X. It's sleek, powerful, and integrates with a dozen platforms. But you're not running a high-volume booking operation. You're managing a few high-touch client relationships.

A frank chat with AI reveals that App X's complexity might actually make your life harder, not easier. If anything, it'll complicate keeping in close touch with your valued clients. You ask your AI partner to compare App X with two simpler alternatives, focusing on ease of use, customization for boutique service providers, and support responsiveness. It turns out the "simple" tool you'd overlooked checks all your boxes. It integrates better with your CRM, fits your workflow, and costs a fraction of what you almost spent.

You just avoided a bad decision **and** you made a *better one*. You also saved yourself hours of trial and error in the process.

AI As The Tool That Knows Tools

The more tools that flood the market, the more valuable it is to have an intelligent guide: *not* to tell you what's best, but to help you reason your way through the noise. This is especially important if:

- You don't have time to test all the latest platforms.
- You've been burned by "shiny object syndrome".
- You need tools that support your style of work, not just your tasks.
- You're managing risk, budget, or team adoption concerns.

AI can be a thinking partner that doesn't have an affiliate link, a bias, or a quarterly sales target. You get insight, structure, and space to think things through at your own pace, in your own way without the hidden agendas.

Getting Started – Step-by-Step

If this is the first time you're considering AI as a thinking partner instead of a tool, you might be wondering: *Where do I even start?*

This is your chance to get hands-on and get some experience. You don't need to be tech-savvy, and you don't need to know how to "prompt engineer." You don't even need a lot of expertise. But getting started is as simple as just diving in and seeing where the process leads you. You just need to understand your business, what specific problems you're trying to solve, and what tangible outcomes you need.

Step 1: Start Talking Out Loud

Begin with something you're wrestling with. Not a task, but an actual question or decision. For example:

- "I'm considering changing up my service model, and I'm trying to decide between two options. I need to serve my existing customers with the same level of care, as well as bring on new customers with a solid service model in place for them. Can you help me explore the tradeoffs? I'd like to implement this over the next quarter, before the holiday season."
- "I had a rough conversation with a formerly loyal client today. They are threatening to go to a competitor, but I don't think they're being

honest about why. Can you help me reflect on what happened and how to respond? I want to keep their business, and I need to understand exactly what the nature of the problem is. Can you help me uncover possible reasons they're threatening to leave, and suggest some possible responses?"

- "Revenue is down this year, compared to last year. My suppliers' prices have gone up, so I think I need to raise my prices, but I'm worried about losing clients. Can you help me think this through?"

These aren't queries you Google. They're questions you'd normally take to a mentor, a coach, or a journal. And they're perfect for establishing a real partnership with AI as a collaborator.

Step 2: Provide Context—And Be Honest

The more context you give, the better the support you'll get. Don't just tell the AI about your business. Tell it where you're stuck. It's okay to say:

- "I'm exhausted, and I'm second-guessing myself."
- "I feel like I should know what to do, but I don't."
- "Here's the emotional weight I'm carrying into this situation."

You're not being dramatic. You're being real. And AI isn't there to dismiss that. It's there to *work with it* as valuable context. If it understands your limitations, it can adapt to your needs. For example, if you tell the AI that you're exhausted and you haven't slept much, it can look more closely for "obvious" things you might be missing. If you share that you're worried and tell it what you're worried about (like making payroll or an insubordinate employee), it can focus specifically on information like cost-cutting ideas, additional revenue streams, or organizational dynamics.

Step 3: Ask It To Help You Reflect, Not Just Answer

This can make a big difference. Instead of asking AI for solutions, ask it for support in your thinking process. Try:

- "Help me map out different options and their possible consequences."

- "Before you answer, can you reflect back what you're hearing in my explanation?"
- "What assumptions might I be making without realizing it? What am I missing?"
- "What questions should I be asking myself right now?"

Think of it like a really good coach: its job isn't to do your thinking for you. It's to help you *do your own thinking more clearly*. In fact, as a solopreneur, having a resource to help you refine your thinking process and reach your own conclusions can pay off in a big way in the long run, as you get in the habit of stopping and thinking things through, validating information, and "getting outside your head" for problem solving.

Step 4: Iterate, Revise, And Practice

If AI's first responses aren't perfect, don't panic. It probably doesn't have enough context to process everything. Add more details, even if they seem minor. Ask follow-up questions. Refine your request. Reframe your goals. This kind of AI leverage isn't a one-shot interaction, it's a conversation.

The more you engage, the more helpful AI becomes. Over time, it can:

- Mirror your voice and values.
- Understand your business rhythms.
- Offer better, faster insights with less prompting.

What starts as clunky becomes fluent. What starts as skeptical becomes indispensable.

Step 5: Use AI When No One Else is Available

Think of AI not as a replacement for human connection, but as an interim support system.

- Is your mentor asleep? AI can talk you through your options and help you weigh pros and cons, until your mentor is available.
- Is your team not ready? AI can help you collect your thoughts or develop a solid plan of action for when they are ready.

- Are you too embarrassed to ask a peer? AI won't judge you for even the "dumbest" questions.
- Or are you just trying to clarify your thinking? AI can sit with you, reflect back, and offer alternatives.

Sometimes that simple presence, that space to think out loud with structure, can make all the difference until you get human help.

Myths, Misuses, and Missed Opportunities

In the midst of the "AI magic," we still have plenty of myths, misuses, and missed opportunities. Companies launch pilots, then discover their data isn't suited for AI integration. "Game-changing" products are marketed aggressively, then don't deliver on their promises of massive ROI. It can be hard to figure out what's signal in the midst of all the noise of overblown promises and obvious hype. And then there are the cautionary tales that we can't afford to overlook, including accounts of people experiencing symptoms of mental illness and even ending their lives after being deeply involved with generative AI chatbots. It's hard to know what's real—and what's just a clever demo with background music and perfect lighting.

Sometimes the best offense is a great defense. Here are some of the most common myths and missteps solopreneurs face when it comes to AI, as well as more grounded, useful ways to think about it.

Myth 1: "AI should give me the right answer right away."

With all the emphasis on streamlining, offloading, and automation, a lot of people expect AI to act like a genius oracle that delivers truth on demand. But if that's your goal with AI, you're setting yourself up for disappointment. AI is not magic. It doesn't "know" the truth, and it's not a "vending machine" for brilliance. While you may get lucky, every now and then, it can't pull the perfect answer out of thin air without detail and context. Plus, it doesn't understand you the way a mentor does. What it *does* do incredibly well is process large patterns of information and reflect structured possibilities back to you.

Instead of:

Using it like Google to come up with simple answers.

Try this:

Approach it as a strategic thought partner. Ask for options, tradeoffs, risks, and reframes. Use AI to *see the landscape,* not mine for one-off facts.

Myth 2: "AI will replace human expertise."

This is one of the most persistent fears and one of the most misunderstood ideas. The goal of using AI isn't to isolate yourself or cut off collaboration. It's to *augment* your thinking, so that when you *do* connect with others, you're more prepared, more centered, and more effective.

Instead of:

Firing your admin, cutting your mentor loose, and ditching your marketing agency.

Try this:

Use AI to prep for meetings, get clear on your objectives, rehearse conversations, analyze feedback, or refine your ideas before involving others. It can make your human interactions *more impactful,* not less.

Myth 3: "I have to be tech-savvy to use AI well."

You don't. But you do need to understand your business. What you need is curiosity, honesty, clarity, and a willingness to talk to it like a thinking partner, not a fact dispenser. The more human you are with it, the more helpful it becomes.

Instead of:

Saying "Give me a content calendar for the next month."

Try this:

Describe your audience, what you're hoping to get them to think/feel/take action on, what topics you'd like to cover, and why. Approach it like a conversation. Don't worry about perfect prompts. Be clear, be real, and engage like you would with a journal or coach.

Misuse: Treating AI like a minion instead of a strategist

It's tempting to use AI just for surface-level tasks: summarize this article, draft this email, generate ideas. Knocking out the low-hanging fruit in a fraction of the time can be absolutely intoxicating. But that barely scratches the surface. What could you do with the time you save? Take advantage of those savings by *bringing it into the deeper parts of your work:* strategic decisions, ideation, product/service development, conflicts, tradeoffs, and more.

Instead of:

Using AI to perform basic tasks like writing emails, call summaries, and marketing materials.

Try this:

Bring it into your planning, debriefing, pricing decisions, hiring processes, emotional processing. Anywhere you think alone, you could be thinking *better* with support.

Missed Opportunity: Never building an ongoing relationship with AI

Many people try AI once or twice, get a mediocre result, and walk away. But like any tool (or any collaborator) it gets better the more you work with it. Over time, it learns your style, your voice, your values. It becomes more relevant, more aligned, and more fluent. And you become more adept at understanding what it does best for you, as well as where to add it to your strategy and operations.

Instead of:

Occasionally doing "drive-by" prompts to get quick answers.

Try this:

Commit to using AI for *your next 10 decisions*. Track how it helps. Refine how you engage with it. Don't judge it by the first five minutes. Give it a chance to become part of your rhythm.

AI isn't just about automation and cost-cutting. It's also here to give you space to think, to plan, to breathe, and to move forward with clarity.

The Bottom Line: You Don't Have to Think Alone Any More

Entrepreneurship has always demanded courage. There's no way around the pressure, the unknowns, and the messy, high-stakes decisions that only *you* can make. But courage doesn't have to mean doing it all by yourself. Fortunately, AI can lighten your load, but not just in terms of offloading marketing copy drafts. It can also help by clarifying your options, holding space for your process, and even pushing back on your assumptions. Whether you're choosing between clients, rewriting your positioning, hiring your first employee, or simply trying to get your feet under you after a tough week, AI can be the partner who sits quietly in the corner, ready to reflect, refine, and respond. On your time. On your terms. And it does it all without ego, without bias, and without fatigue. It's not the loudest voice in the room. It's not a magical oracle. It's simply a presence to help you hear yourself think. And sometimes that's the most important help of all.

So take the leap. Start the conversation. Build the habit. In a world where change is constant and the stakes are higher than ever, thinking well can be your greatest competitive advantage.

And now, you don't have to do it alone.

About the Author

This author is a seasoned technology veteran with over 3 decades experience in designing and building information solutions for startups and the enterprise. She has been working with the building blocks of AI since 2000, and has led search-based initiatives for multi-million customer bases since the end of the 20th century. She now specializes in "relational AI" with an emphasis on empowering entrepreneurs and solo operators through cutting-edge AI implementations.

Chapter 15

Who Holds the Wheel

Power, Cost, and Care in Automotive AI

By Elizabeth Eagle-Simbeye

CARS ARE NO LONGER just machines that move us from one place to another. With AI embedded across every stage of the journey, they are becoming platforms that connect us, anticipate our needs, and quietly reshape what value, ownership, and freedom mean for all of us who build and use them.

If you work in AI or are trying to break into it, you probably already know that automotive is one of the loudest stages for all of this. It is where safety, data, privacy, money, and risk come together in very public ways.

It is also where you can feel the gap between how products are described in leadership decks and how they land in real lives, especially for women who drive late at night, who manage school runs, or who have been taught to be careful in a way men often are not.

In this chapter, I look at how AI is transforming the automotive experience, from that first buying journey to the everyday reality of living with a connected car. I anchor it in one memory: a car showroom visit with my mum, where safety, affordability, and reliability were the deciding factors. Those same questions remain. The difference is that the answers now come through algorithms, digital platforms, and connected services that people like us help design, ship, and defend.

A Showroom Lesson

I still remember the first time I really noticed what it meant to buy a car. I was a teenager, sitting beside my mum, who was in her late thirties at the time. She wanted something safe, affordable, and reliable. Fuel consumption mattered too, because every pound counted once you filled the tank week after week.

I, on the other hand, was distracted by the smell of new upholstery, the satisfying thud of doors closing across the showroom, and the fantasy of test-driving something with a powerful stereo and leather seats. I remember drifting toward a metallic blue model that sat slightly apart from the rest, the kind designed to make teenagers believe adulthood came with sunroofs and speakers that rattled the mirrors. I traced the curve of the bodywork with my eyes and imagined myself behind the wheel, music loud enough to feel like a personality. It felt like possibility–like freedom wrapped in metal.

Meanwhile, my mum stayed rooted in the practicalities. She moved methodically from one car to the next, checking boot space for grocery trips, asking about safety ratings, and trying to understand the real monthly payments beneath the glossy brochures. She circled back to the same questions again and again. The salesman circled around them with equal persistence, answering the questions he preferred.

Looking back, I realise that day was not only about choosing a car. It was the moment I understood that two people can stand in the same space, look at the same object, and see entirely different futures reflected back at them.

She was scanning for the things that made life safer and more manageable. I was imagining freedom, identity, and the feeling of movement.

What stuck with me was not the car. It was the imbalance. Information flowed one way.

The salesman knew the margins, the flexibility, the true floor of the deal. My mum knew her budget and her limits. She kept circling back to the same questions. He deflected toward the ones he preferred.

If you have ever sat in a product review where the narrative is polished and the data is hand-picked, you know that feeling. The story sounds confident. The gaps sit in the white space and in the assumptions.

The showroom has mostly moved online now. Many buyers never visit one in person. They compare models, explore options, and arrange finance from the sofa before scheduling delivery to their door.

Convenience, personalisation, and visible pricing have replaced the handshake as the currency of trust.

My mum's concerns are still at the centre. Safety now involves advanced driver assistance powered by AI. Affordability has expanded to include subscriptions that unlock digital features on top of the purchase price.

Fuel consumption has become a conversation about electric charging, battery life, and the systems that optimise both in the background.

Reliability, once judged by whether an engine would start on a cold morning, now includes whether software behaves predictably when conditions fall outside the norm.

Cars have become connected platforms, always updating and always collecting data. Some of that intelligence genuinely improves safety and comfort. Some of it introduces new points of failure. A model that performs well most of the time can still fail spectacularly at the edges. And when it does, the impact is immediate and physical.

Cars have become connected platforms, always updating and always collecting data. Yet the same questions remain.

Who holds the information? Who holds the power?

The Algorithmic Showroom

The modern showroom lives in the browser, but it is no longer passive. Instead of walking the lot and assessing each model up close, people scroll through curated recommendations and filtered results. The journey is quieter, cleaner, and far more controlled by algorithms than by weather or instinct.

On the surface, this looks more transparent. Prices appear in clear text. Finance quotes show up instantly. Nobody waits for a salesman to return with another sheet of numbers.

The theatre has not disappeared. It has changed rooms. AI now decides what people see when they search. Results are not neutral. They are shaped by algorithms that optimise for sales, stock levels, margins, and sometimes predicted lifestyles.

A family researching child seats gets nudged toward larger cars. A driver in a low-income postcode may be shown fewer premium models. A young woman browsing at night may be placed in a different bucket altogether.

If you work in AI, you already know how easily a model learns the wrong lesson. A small skew in the distribution becomes a confident assumption. That assumption becomes the default ranking. People think they are browsing freely when they are being softly guided within the boundaries the model drew for them.

This is where responsibility lives. A digital showroom can either replicate the old power imbalance with cleaner typography, or it can give customers enough clarity to feel in control. The difference lies in what we choose to optimise for. Margin is easy. Long-term trust takes more work.

Car dealers and manufacturers are rarely rewarded for transparency. They are rewarded for moving stock, upselling features, and protecting margin. Trust pays off more slowly, through repeat customers, longer ownership, and quieter loyalty. Those outcomes matter, but they are harder to measure and easier to deprioritise.

A more useful showroom is still possible. Imagine if drivers could see realistic long-term costs based on how they actually travel. Charging expenses, insurance bands, maintenance patterns, and resale value surfaced as clearly as colours and trims. That is affordability made visible. Not theatre. Not persuasion. Actual support.

Buyers also have a role. When opaque pricing and buried subscriptions are rewarded with purchases, those patterns persist. When people ask harder questions, compare true cost of ownership, or walk away from experiences that feel misleading, transparency becomes a competitive advantage rather than a moral extra.

Women who shop for cars often carry safety, cost, and reliability as their primary concerns. A thoughtful AI could meet those concerns directly. A careless AI could bury them under optimisation strategies that prioritise margin over meaning.

As builders, we sit in the middle. We may not control the incentives, but we shape what is surfaced, what is hidden, and what becomes normal. Over time, those choices decide whether trust is optional or expected.

Ownership Redefined

When my mum bought her car, ownership meant exactly what it said. A set of keys, a logbook, and a long-term relationship with a single machine. The car was part of the family budget, not a subscription stack.

Ownership has evolved into access. Drivers subscribe, lease, swap, and share. They pay monthly for flexibility, which sounds smart until you add up the total. Many people now spend more over three years than my mum spent across a decade.

For some of us, this new flexibility is genuinely useful. Lives shift quickly. Jobs change. Family responsibilities grow or shrink. Safety needs shift. And predictable costs can be comforting. A higher monthly payment can feel worth it if it removes the risk of a sudden repair bill that wipes out savings. Many subscription or lease models include maintenance plans, repairs, and early swap options. For some drivers, that coverage feels like peace of mind baked into the contract, not a luxury.

The cost comes quietly. Monthly payments feel manageable. They create a sense of control. But add them together across three or five years, and the number tells a different story. Subscriptions spread out the pain, which is why companies love them. They also obscure the truth of what ownership now costs.

Electric vehicles add more layers. Range, charging access, and battery health become daily concerns. AI now manages much of this in the background. Charging is scheduled when rates are low. The battery is protected from overuse. The route is chosen to make sure you never arrive at a dark car park with one percent left.

This is helpful, but not neutral. When a car decides how you drive and when you charge, you surrender a degree of agency. For those who already make constant safety calculations about time, place, and visibility, giving that control to a black box can feel uncomfortable.

Some transparency exists today, but it is uneven. A few systems explain why charging was scheduled at a certain time or how a route was chosen. Others offer simple toggles for cost savings or battery preservation. However, the details are often buried, inconsistent, or treated as optional rather than essential.

True transparency would turn that discomfort into trust.

- Clear explanations in plain language.
- Predictable behaviour that doesn't require guesswork.
- Settings that make it easy to override automation, rather than hunting through menus.

Ownership is changing, but agency should not be collateral damage.

Under the Bonnet: How AI Actually Works

When my mum asked about engines, she wanted to understand what could break, how expensive the repair might be, and whether this model had a reputation for behaving in winter. Today, when people ask what is under the bonnet, they often mean something else entirely. They want to know what the AI is doing.

Modern cars collect data constantly. Cameras, radar, ultrasonic sensors, accelerometers, and GPRS. Every turn of the wheel feeds machine learning systems that try to recognise patterns. Deep neural networks become the second brain beside the engine. They try to understand if the dark shape by the curb is a cyclist, a bin, or a child stepping backwards.

If you work with data, you understand the vulnerable parts of this story. Data is not neutral. It never has been. If the model is trained mostly on daylight conditions, it performs worse at night. If the training set skews toward certain postures, genders or heights, detection accuracy drops for smaller bodies. If weather conditions from certain regions are underrepresented, the model makes riskier guesses when confronted with them.

Some failures are trade-offs. A system designed to react quickly may overreact. A system tuned for caution may misread a harmless bump as a collision. These moments make headlines not only because the stakes are high, but because the gap between intention and outcome becomes painfully visible. The same mechanism that prevents one crash can create panic in another. For a driver, both experiences feel equally real.

The auto industry likes to talk about scale. Millions of miles. Billions of frames. Yet none of that matters if the edge cases involve the exact people who are already overlooked.

- Women walking with prams.
- Older adults moving slowly.
- Cyclists carrying shopping.
- People navigating roads while managing children.

These small variations matter because they exist everywhere.

The leap of faith required from drivers grows every year. My mum worried about whether her car would start on a frosty morning. Drivers today must trust that a machine brain will make the right call in a moment that leaves no room for error. And they must trust that the car will not *overreact* in a way that creates a new kind of danger. That trust must be earned, not assumed.

Patterns, Practice, and Generative AI

Machine learning is a practice made permanent. The model watches patterns and then assumes they are rules. Sometimes that helps. Sometimes it traps people in narratives they did not choose.

If you brake hard once to avoid a dog, that can label you as a harsh driver. If you pause at junctions because visibility is poor, the model can misread that as hesitation. Behaviour becomes data. Data becomes judgment. The driver becomes a score.

For some of us, this is familiar in a different context. Our behaviour is often judged without the surrounding story. AI can amplify that pattern if the designers do not build room for context or override.

Generative AI introduces new forms of capability. Simulated scenarios help train autonomous systems. Synthetic data fills gaps in rarely-seen conditions. Assistants generate spoken guidance. Engineers can test safety logic against millions of hypothetical cases without risking lives. These are real advances.

The hazard sits in confidence. Generative models produce output that sounds sure, even when the reasoning behind it is thin. In a chat app, that is annoying. In a vehicle, it is dangerous. A model that hallucinates instructions or mislabels a scenario moves risk from the edge to the centre.

This is where we can lead differently. Designers, engineers, and product teams make choices drivers cannot. We can ask grounding questions.

- What does the driver need to know?
- What can they safely ignore?
- Where does ambiguity live?
- What happens if something fails at the worst possible moment?

These are design questions. They are also safety questions. Drivers and buyers rarely see those choices. They experience the outcome, not the logic behind it. They can decide whether to trust a system, but not how that trust was constructed.

A good system acknowledges uncertainty. A bad one hides it. Our responsibility is to build systems that are honest about their limits, especially when the people relying on them cannot see what sits underneath.

In-Car Intelligence: From Gimmick to Infrastructure

When my mum drove, intelligence meant reading the road, listening for engine noise, and keeping one eye on the radio. Safety depended on awareness, not software.

Inside today's vehicles, intelligence sits everywhere. Voice assistants, route prediction, climate control, or predictive maintenance. Some of it is genuinely helpful. Some of it exists because someone needed to justify a budget line.

Useful intelligence removes cognitive load.

- Automatic maintenance alerts.
- Navigation that adapts to real traffic.
- Charging suggestions that save money.
- Tasks disappear quietly.

Gimmicks interrupt.

- A chirpy offer for a nearby cafe when you are driving home alone at night.
- A message that fills half the screen at the wrong moment.
- A recommendation that has nothing to do with safety or context.

However, context carries weight.

- A badly timed prompt can distract at the exact moment you need to stay alert.
- A complicated menu can slow you down in a dark car park.
- A broken voice assistant can leave you without hands-free navigation during stressful routes.

This is why I test every in-car feature with one question. Does it disappear when it is not needed? If a system demands attention every few minutes, it is not intelligent. It is needy.

Thoughtful design acts like a good co-pilot. Quiet. Observant. Supportive. Present only when required.

Efficiency, Choice, and the Quiet Co-Pilot

My mum measured efficiency in petrol money. She filled the tank, noted how long it lasted, and adjusted accordingly. Efficiency lived in her routine, not the dashboard.

AI has turned efficiency into a constant flow of analysis. Predictive maintenance. Real-time traffic. Charging optimisation. Eco driving suggestions. Some of this is genuinely empowering. Nobody wants a breakdown that could have been avoided. Nobody wants to pay higher costs when a cheaper option exists.

The challenge is tone and autonomy. Some drivers feel supported by efficiency nudges. Others feel judged or micromanaged. This difference is especially visible when we choose a longer or better-lit route at night, or avoid isolated charging spots even when they are more efficient. The map may think it is helping. The driver knows their reality better.

Design must respect that. Efficiency should serve the person, not command them. A system that explains trade-offs and lets you choose freely builds trust. A system that nags or penalises builds resentment.

Good design supports the reasons behind your choice, even when they are invisible to the algorithm.

The Basics Still Matter

Automotive brands love to talk about autonomy levels, AI-powered features, and futuristic dashboards. Customers talk about something else. They talk about whether the car starts reliably. They talk about whether the system connects to their phone. They talk about whether the screen freezes at the wrong moment.

Reliability, clarity, and responsiveness matter more than any feature launch. A car that does fewer things well feels more advanced than a car full of features that do not work.

Drivers often ask the right questions. They notice friction immediately. But it is builders who have the leverage to act on those questions at scale. This is where designers, engineers, and product teams can bring crucial leadership. We ask about failure modes. We plan for what happens when the network drops. We test whether the screen is still readable with cold hands or in bright sunlight. These questions are not glamorous, but they are the difference between confidence and frustration.

Innovation does not matter if the basics break. Safety and trust live in the foundations, not the headline.

Cost, Value, and the Price of Digital Extras

Affordability used to be simple. My mum added up the purchase price, insurance, fuel, and repairs. That was it.

Today, the sticker price is only the start. Many features now require monthly payments: heated seats, advanced navigation, driver assistance, and premium connectivity. A car can feel like a streaming service on wheels.

Subscriptions can be fair when they reflect real value. Rapid charging access makes sense as an add-on. So does specialised navigation. Paying to unlock hardware that is already installed feels different. It feels like the company is charging you for access to something you already own.

AI could actually fix this. It could show the true cost of ownership based on usage patterns instead of marketing bundles. It could explain whether a subscription is worth it based on your driving, not someone else's assumptions. But it only does this if the company chooses transparency over revenue smoothing.

As builders, we often have more influence than we realise. We decide how costs are presented. We decide whether a customer sees the truth or the theatre.

Loyalty in an AI-Driven Ecosystem

My mum returned to brands that treated her fairly. Loyalty was earned through safety, reliability, and honesty. Simple.

Now loyalty lives across hardware, apps, services, and support. A driver might love the car but hate the app. Or they might tolerate a few hardware flaws because the digital experience feels calm and well-designed.

Today, Loyalty now rests on three pillars: safety, value, and ease.

- **Safety** includes physical protection and trust in systems that behave reliably under pressure.
- **Value** means paying for something that feels worth the cost over time, not just at the point of purchase.

- **Ease** means every small task feels manageable rather than irritating.

People rarely talk about algorithms when they describe loyalty. They say the car just works. Or they say it gets in the way. That is AI by another name.

For those working in AI and design, this is familiar territory. We are often the ones speaking up when a friction point is dismissed as minor. We know minor friction accumulates. We also know it is often felt most intensely by the people who are already carrying a mental load.

Challenges on the Road Ahead

Every layer of AI promises progress. It also opens new problems.

Unpredictability persists. Models trained on patterns struggle with rare or unusual events, which are often the exact moments that matter most.

Data volume overwhelms even sophisticated systems. Millions of signals compete for space. Designers must decide what to show, what to hide, and how to maintain clarity when everything feels urgent.

Insurance models are shifting. Behavioural telemetry rewards some drivers and penalises others. Without transparency, people can be judged by data that lacks context or nuance.

Privacy is a growing concern. Cars now monitor alertness, behaviour, and sometimes biometrics. Connecting a mobile phone to an infotainment system can expose personal data far beyond navigation or music, including contacts, messages, and location history. In many countries, this data sits in regulatory grey zones, protected unevenly or not at all once it moves between devices, vehicles, and third-party services.

This data can prevent harm. It can also turn ordinary actions into information that is stored, transmitted, or sold. So, the question is no longer just *what* the system sees, but whether the driver ever had a chance to agree to that level of visibility.

A system that watches to keep you safe can still feel intrusive. Consent matters. Choice matters. Drivers deserve to know what is being measured, what is retained, and what agency they have over it.

That emotional truth must be acknowledged in design, because **safety without consent is not safety—it is surveillance dressed as support.**

Rethinking Value on the Road Ahead

When I think back to sitting with my mum in that showroom, I remember how focused she was on the essentials. Would the car protect her? Could she afford to run it? Would it carry us reliably through the week?

Those questions shaped the way she moved through that space, and they stayed with me long after I stopped trailing behind her between rows of cars.

Now I am the one asking them.

The landscape has changed dramatically. Cars function as connected platforms. AI shapes navigation, safety, cost, and comfort. The basics my mum cared about have become more complex, layered beneath software decisions and automated systems that follow me long after I drive away from the dealership. Yet the fundamentals still matter most.

The promise of AI is real. It can reduce friction, prevent accidents, and lower running costs. But buying a car today means understanding the trade-offs hidden inside those promises. A feature that claims efficiency may reduce control. A personalised setting may rely on data I didn't intend to share. An automated decision can feel helpful in theory and disorienting in the middle of real life. Researching a car now is less about reading specifications and more about imagining how these invisible systems will behave in the rhythm of *my* everyday routines.

As a buyer, the questions I need to ask have expanded. What am I paying for? Who controls the data? How predictable is this system when conditions shift? And when convenience is offered, what is the cost of accepting it?

As someone working in this industry, the responsibility sits even heavier. I help shape the systems that answer those questions. I push for clarity, fairness, and care. I'm aware of the edges others overlook. I design for the moments where safety and humanity matter more than speed. And I carry that awareness with me not just as a practitioner, but as a driver who relies on these systems myself.

If my mum were with me today, walking around a dealership or scrolling through online models, her questions would be unchanged: Will it keep us safe? Can we afford it? Will it make life easier? I find myself asking the same, just with new layers behind each one.

The future of mobility will not be judged by how autonomous it becomes. It will be judged by whether AI supports the people who rely on it, with care, honesty, and respect. If we build that future—and if I choose wisely as a driver within it—the road ahead will feel more human, not less.

About the Author

***Elizabeth Eagle-Simbeye** is a Principal UX Consultant and Design Strategist based in Manchester, UK. She has spent over two decades guiding digital transformation across industries such as automotive, finance, and retail. Her work focuses on how technology, including AI, shapes not only the products we use but also the culture of the organisations that build them. She mentors designers and writes on leadership and design culture, always grounding complex topics in human terms.*

Chapter 16

AI and Agriculture

Feeding the Future Through Intelligent Farming

By Farida Khalaf

1. Introduction: The Urgency of Agricultural Transformation

THE CONVERGENCE of artificial intelligence and agriculture represents one of the most critical technological transformations of our time. When I examine the current state of global food systems, struggling to feed a growing population while grappling with climate change and resource constraints, I become convinced that AI is not just supplementing farming practices, but fundamentally redefining how we must grow food to survive as a civilization.

The mathematical reality is stark: The global population is projected to reach around 9.7–10 billion by 2050, placing sustained pressure on food systems and requiring significant growth in agricultural production (OECD & FAO, 2025). Climate change is amplifying weather variability and driving a sharp rise in climate-related disasters; weather-related events nearly doubled between 2000 and 2019, and associated economic losses now exceed USD 2.3 trillion per year when cascading impacts are included (UNDRR, 2020; UNDRR, 2025). Agriculture uses about 70 percent of global freshwater withdrawals and agrifood systems account for roughly 30 percent of anthropogenic greenhouse gas emissions, underscoring that expansion of traditional farming practices is environmentally unsustainable (FAO, 2025).

This is where AI becomes absolutely essential, not merely helpful. The technology represents our most viable path for achieving what seems impossible: dramatically increasing food production while simultaneously reducing environmental impact. But let me be clear about what AI is not: it's not a magical solution that will solve agriculture's problems overnight. It's a sophisticated toolset that, when properly implemented, can transform farming from an intuition-based practice into a precision science.

Having researched AI applications across multiple industries, agriculture strikes me as uniquely positioned to benefit from intelligent systems. Unlike sectors where AI often replaces human decision-making, in agriculture, AI enhances and amplifies human expertise. A farmer's knowledge of their land, accumulated over generations, combined with AI's ability to process vast datasets and identify patterns, creates a synergy that neither could achieve alone.

However, analysis by SDG (Carcamo et al., 2023) reveals significant barriers to AI adoption in agriculture, from the digital divide that separates large commercial operations from smallholder farmers, to legitimate concerns about data ownership and technological dependence. These challenges don't invalidate AI's potential; they demand that we implement these technologies thoughtfully and equitably.

This chapter explores how AI is revolutionizing agriculture across the entire value chain, from soil analysis to supply chain optimization. More importantly, it examines the real-world impact of these technologies and the critical decisions we must make to ensure AI serves all farmers, not just the well-resourced few. The future of food security depends on getting this right.

2. The AI Agricultural Revolution: Beyond Simple Automation

Artificial intelligence in agriculture represents a fundamental shift from reactive to predictive farming. Traditional agriculture has always been about responding to problems after they occur, treating diseases once symptoms appear, irrigating when crops show stress, harvesting when schedules dictate. AI transforms this paradigm by enabling farmers to anticipate and prevent problems before they impact yields.

Pre-Season Intelligence: Planning with Data

Modern AI-powered farm management begins months before planting. Machine learning algorithms analyze historical data of soil composition, weather patterns, commodity prices, and yield outcomes to recommend optimal crop selection and field allocation. This goes beyond simple data analysis to predictive modeling that can simulate multiple scenarios to identify the most promising strategies.

Research from the University of Illinois demonstrates that farms using AI-driven crop planning achieve measurably higher profitability compared to traditional decision-making approaches, though exact percentages vary significantly based on farm size, location, and implementation quality (Khanna et al., 2021). The key advantage lies not in guaranteed returns but in risk reduction through better-informed decisions.

Real-Time Field Intelligence

During the growing season, AI systems continuously monitor crop health through multiple data streams. Internet of Things (IoT) sensors embedded throughout fields collect real-time data on soil moisture, temperature, pH levels, and nutrient availability. While IoT devices have long collected this data, AI now plays a crucial role in analyzing and acting on it. AI algorithms process the real-time data to detect patterns, predict issues like irrigation needs or disease outbreaks, and provide actionable insights. Satellite imagery and drone surveillance further enhance this by offering visual data on crop development, stress indicators, and pest pressure. This combination of IoT and AI helps farmers make more precise, data-driven decisions throughout the growing season.

The sophistication of this monitoring continues to improve. Current AI systems can detect plant stress 7-10 days before it becomes visible to human observers. Computer vision algorithms trained on large plant image datasets can differentiate between multiple types of leaf diseases with accuracy rates exceeding 90%, performance that matches or exceeds trained agronomists in controlled studies (Saleem et al., 2019).

This real-time monitoring capability represents a revolutionary departure from traditional agricultural practice. Where farmers once relied on periodic soil tests and visual inspections conducted weeks apart, today's AI-powered systems provide continuous, hyperlocal analysis of soil conditions, plant health, and crop development. IoT sensors can detect subtle changes in soil moisture, nutrient levels, and pH that occur within hours rather than days. Satellite imagery refreshed multiple times per week reveals stress patterns invisible to ground-based observation. Mobile diagnostics apps powered by computer vision can identify plant diseases from smartphone photos with accuracy that rivals trained extension agents.

Precision Application: The End of Blanket Treatments

One of AI's most transformative applications involves enabling precision application of inputs. Traditional farming applies water, fertilizers, and pesticides uniformly across entire fields. AI-powered systems recognize that every section of farmland has unique characteristics and needs.

Variable-rate technology guided by AI can adjust seed spacing, fertilizer application, and irrigation intensity in real-time as equipment moves across fields. Academic studies document substantial reductions in input use, reporting nitrogen fertilizer reductions of 15-25% while maintaining yields are consistently reported, with pesticide applications reduced by 70-90% through targeted spot-treatment (Zhang et al., 2020).

Autonomous Operations: Machines That Think

The emergence of autonomous farm equipment represents a quantum leap in agricultural capability. These systems go beyond GPS-guided tractors following pre-programmed routes; they're intelligent machines capable of complex independent decision-making.

Modern autonomous systems can detect obstacles, adjust operations based on soil conditions, and modify routes to minimize soil compaction. More significantly, they operate continuously, 24 hours per day during critical windows like planting and harvesting when timing determines success or failure.

This level of automation will be crucial for addressing labor shortages in agriculture, which have reached crisis levels in many regions, such as California's Central Valley, Florida's citrus industry, and parts of the Midwest in the USA, as well as Italy, Brazil, Australia, and Spain. However, we must acknowledge that autonomous systems also raise important questions about employment in rural communities and the skills farmers will need in an increasingly automated industry.

Corporate Consolidation and the Data Revolution

The transformation of agriculture through real-time monitoring and AI-powered precision has created unprecedented opportunities and concerning market dynamics. The 2018 acquisition of Monsanto by Bayer exemplifies how traditional seed and chemical companies have evolved into comprehensive agricultural data platforms. Bayer's Climate FieldView system now monitors millions of acres globally, collecting continuous streams of data about soil conditions, crop performance, and management practices.

This shift represents more than technological advancement; it fundamentally changes the relationship between farmers and agricultural suppliers. Where farmers once purchased seeds and chemicals seasonally, they now stay involved with AI platforms and suppliers year-round due to the actionable data they continuously generate. Unlike traditional inputs, this data becomes more valuable over time, providing companies with insights to develop new products, optimize supply chains, and understand market dynamics.

The concentration of agricultural data within a small number of technology companies raises fundamental questions about farmer autonomy and market competition. Real-time soil, plant, and crop analysis generates detailed intelligence about agricultural operations that was impossible to capture through traditional periodic sampling. While these capabilities offer tremendous benefits for precision agriculture, they also concentrate unparalleled agricultural intelligence within corporate databases, creating new forms of dependence that extend far beyond seasonal input purchases.

3. Crop Yield Optimization: The Science of Precision Agriculture

AI's promise in agriculture centers on a straightforward proposition: grow more food on the same land while using fewer resources. Achieving this requires unprecedented precision in every aspect of crop management, from soil preparation to harvest timing.

Soil Health: The Foundation of Intelligent Farming

Healthy soil forms the cornerstone of productive agriculture, yet traditional soil analysis relies on periodic lab tests that provide limited insights into dynamic conditions. AI-powered soil monitoring systems continuously track multiple parameters, moisture content, nutrient levels, organic matter, pH, and microbial activity.

Machine learning models process this data to create detailed soil health maps that guide precision management decisions. These systems can predict nutrient deficiencies weeks before plants show symptoms and recommend optimal fertilizer formulations for specific field zones.

Studies from major agricultural universities document measurable benefits: fertilizer cost reductions of 15-30% combined with yield improvements of 8-12% are commonly reported in peer-reviewed research (Giller, 2021). More importantly, this precision reduces nutrient runoff, addressing a major contributor to water pollution.

Water Management: Conservation Through Intelligence

Water scarcity increasingly constrains agricultural production worldwide, with groundwater depletion accelerating in major farming regions. AI-powered irrigation systems represent our most promising tool for maintaining productivity while dramatically reducing water consumption.

These systems integrate real-time soil moisture data, weather forecasts, crop growth models, and evapotranspiration calculations to deliver precise amounts of water exactly when and where crops need it. Machine learning algorithms continuously refine irrigation schedules based on crop response and environmental conditions.

The documented results are substantial. Precision irrigation systems achieve water savings of 20-40% while maintaining yields, according to multiple field studies (Jones et al., 2021). In drought-prone regions like California's Central Valley, this technology literally keeps farms viable during extended dry periods.

Weather Forecasting: Hyperlocal Precision

Agriculture has always been vulnerable to weather, but AI is changing this relationship by providing hyperlocal forecasts with improved accuracy. Machine learning models trained on historical weather data, satellite imagery, and real-time sensor networks can predict conditions at the field level with remarkable precision.

These forecasts enable proactive management decisions: adjusting irrigation before drought stress occurs, applying protective treatments before severe weather, and optimizing harvest timing to avoid weather-related losses. The ability to anticipate and prepare for weather events represents a fundamental improvement in agricultural risk management.

What strikes me most about these precision agriculture applications is their cumulative effect. Individual improvements might seem modest, but when combined across entire farming operations, they transform the economics and environmental impact of food production. The question is whether we can scale these benefits beyond well-resourced commercial operations to the smallholder farmers who produce most of the world's food in developing countries (FAO, 2021).

4. Disease and Pest Management: Early Detection Saves Crops

Crop diseases and pest infestations cause agricultural losses exceeding $220 billion annually worldwide, losses that could theoretically feed 600 million people (Oerke, 2006). Traditional pest management relies on scheduled treatments and reactive responses to visible damage. AI transforms this approach through early detection and predictive intervention strategies.

Computer Vision: Seeing What Humans Cannot

The most visible application of AI in pest management involves computer vision systems that identify diseases, pests, and nutrient deficiencies from plant images. These systems, accessible through smartphone apps or deployed on drones, democratize expert-level crop diagnosis.

Current commercial systems can identify over 200 different plant diseases and pests with accuracy rates typically ranging from 85-95% in real-world conditions, according to validation studies (Mohanty et al., 2016). More importantly, they detect problems 3-7 days earlier than visual inspection, providing critical time for intervention before significant damage occurs.

I find the developments in hyperspectral imaging particularly compelling. AI analyzes light wavelengths invisible to human vision to detect plant stress. This technology can identify viral infections in crops before any visible symptoms appear, a capability that could prevent entire fields from being lost to disease.

Predictive Modeling: Forecasting Outbreaks

Beyond detection, AI enables prediction of disease and pest outbreaks based on environmental conditions, crop development stages, and historical patterns. Machine learning models trained on agricultural databases can forecast pest pressure weeks in advance, allowing farmers to implement preventive measures.

These predictive capabilities prove particularly valuable for managing devastating pests like fall armyworm, which can destroy entire maize crops within days. Early warning systems powered by AI give farmers time to apply biological controls or deploy resistant varieties before infestations become unmanageable.

Precision Treatment: Targeted Interventions

AI-guided treatment systems represent the future of sustainable pest management. Instead of applying pesticides across entire fields, smart spraying systems identify individual pests or diseased plants and treat only those areas requiring intervention.

The Blue River Technology "See & Spray" system, now owned by John Deere, exemplifies this approach. Independent field trials document herbicide reductions of 77-90% while maintaining effective weed control (Gerhards & Oebel, 2006). Similar systems under development for fungicide and insecticide applications promise comparable reductions in chemical inputs.

This precision approach addresses growing concerns about pesticide resistance, environmental contamination, and impact on beneficial insects. By using chemicals only where and when necessary, AI-guided systems extend the effective life of existing pest control tools while reducing environmental impact.

What concerns me about current pest management AI is the concentration of these technologies among large agribusiness companies. The farmers who most need early disease detection, smallholders in developing countries with limited access to extension services, often cannot afford or access these systems. This represents both a moral failing and a missed opportunity for global food security improvement.

5. Autonomous Farm Equipment: The Rise of Intelligent Machines

The transformation of farm equipment from mechanical tools to intelligent machines represents agriculture's most dramatic technological leap. Modern autonomous systems don't simply follow pre-programmed instructions; they perceive their environment, make complex decisions, and adapt their operations in real-time.

Smart Tractors: Beyond GPS Guidance

Today's autonomous tractors represent sophisticated AI systems capable of complex decision-making. Equipped with computer vision, LIDAR (**LI**ght **D**etection **A**nd **R**anging), and advanced sensors, these machines can navigate fields, avoid obstacles, and adjust their operations based on soil conditions and crop requirements. LIDAR uses laser light to measure distances and create detailed 3D maps of the environment, which helps the tractors "see" and react to their surroundings.

Modern autonomous tractors can plant seeds with centimeter-level precision, adjust seeding rates based on soil conditions, and modify their routes to minimize soil compaction, all while operating continuously during optimal weather windows. The productivity gains are measurable: autonomous equipment can operate 24 hours per day during critical periods, increasing operational capacity by 30-50% compared to human-operated machinery (Lowenberg-DeBoer et al., 2020).

Autonomous Treatment Systems: Precision Agriculture in Action

Some of the most impressive agricultural AI applications involve targeted treatment systems that identify and address specific problems with surgical precision. These systems combine computer vision, GPS guidance, and precision application equipment to deliver treatments exactly where needed.

Robotic weed control systems exemplify this capability. Using computer vision to distinguish between crops and weeds, these machines can mechanically remove or precisely spray individual weeds while leaving crops untouched. The herbicide savings are dramatic, often exceeding 90% compared to broadcast applications (Fennimore & Cutulle, 2019).

Integration Challenges and Realities

Despite impressive capabilities, autonomous farm systems face significant implementation challenges. High capital costs, often $500,000 or more for complete systems, limit adoption to large commercial operations. Technical complexity requires new skills and support infrastructure that many farming communities lack.

More concerning is the potential for increased technological dependence. Farmers using autonomous systems must rely on manufacturers for software updates, technical support, and equipment maintenance. This dependence raises questions about operational resilience and farmer autonomy that the agricultural industry must address.

However, solutions are emerging to broaden access to autonomous farming technologies. Equipment-as-a-service models, cooperative ownership structures, and government support programs are beginning to make these technologies accessible to smaller operations. The key is ensuring that automation serves farmers rather than creating new forms of dependence.

6. Supply Chain Intelligence: From Farm to Fork

Agricultural production represents only one component of the food system. The journey from farm to consumer involves complex supply chains where inefficiencies generate massive food waste, economic losses, and environmental impacts. AI transforms these supply chains through intelligent coordination, predictive analytics, and optimization systems.

Post-Harvest Intelligence: Preserving Value

Post-harvest losses account for approximately 14% of global food production, representing both economic waste and missed opportunities to address food insecurity (FAO, 2019). AI-powered storage and handling systems address these losses through intelligent monitoring and control systems.

Smart storage facilities equipped with AI sensors continuously monitor temperature, humidity, air quality, and pest activity to maintain optimal conditions for different commodities. Machine learning algorithms predict spoilage risks and recommend interventions before products deteriorate.

Commercial implementations document substantial improvements. Cold storage facilities using AI management systems report 30-40% reductions in spoilage compared to conventional operations (Kumar & Kalita, 2017). For perishable crops like fruits and vegetables, this improvement often determines profitability.

Demand Forecasting: Aligning Supply and Markets

AI's most valuable contribution to agricultural supply chains involves demand forecasting systems that predict consumer requirements with improved accuracy. These systems analyze historical sales data, economic indicators, weather patterns, and social media trends to forecast demand for specific commodities.

Accurate demand forecasting enables farmers to make informed planting decisions, reducing the boom-and-bust cycles that characterize many agricultural markets. Retailers can optimize inventory levels, reducing waste from overstocking while avoiding shortages that drive up consumer prices.

Major food distributors using AI-powered demand forecasts report accuracy improvements of 10-20% compared to traditional methods, translating directly to reduced waste and improved profitability throughout the supply chain (McKinsey Global Institute, 2020).

Consumer Impact: How AI Agriculture Affects Your Grocery Bill

Here's what many people don't realize about AI in agriculture: it's already affecting what you pay for food and what options are available in your local grocery store. When AI systems optimize crop yields and reduce waste, those efficiency gains eventually translate to more stable food prices for consumers.

AI-powered supply chain optimization means fresher produce reaches stores faster, extending shelf life and reducing the markup retailers need to cover spoilage losses. Precision agriculture that reduces input costs (such as less fertilizer, pesticide, and water) helps keep production costs manageable even as other expenses rise.

The flipside concerns me: if these AI benefits remain concentrated among large agribusiness operations, we could see increased market concentration and reduced competition, potentially leading to higher consumer prices despite lower production costs. Ensuring broad access to agricultural AI technologies isn't just about fairness to farmers, it's about maintaining competitive food markets that benefit consumers.

7. Benefits and Transformative Opportunities

AI integration into agriculture offers benefits extending far beyond individual farms or agribusinesses. These technologies represent our most promising pathway to achieving global food security while addressing environmental sustainability and social equity challenges.

Resource Efficiency: Quantified Environmental Benefits

AI's most immediate benefit involves measurable improvements in resource use efficiency. Meta-analyses of precision agriculture studies consistently document water consumption reductions of 20-30%, fertilizer use decreases of 15-25%, and pesticide application cuts of 50-80% while maintaining or improving yields (Khanna et al., 2021).

These efficiency gains address agriculture's most pressing environmental challenges. Reduced fertilizer use decreases water pollution and nitrogen runoff, a primary cause of coastal dead zones that affect fisheries worldwide. Lower pesticide application protects beneficial insects and reduces human exposure to harmful chemicals. Improved water efficiency helps preserve groundwater resources essential for long-term agricultural viability.

The economic implications are equally significant. Input costs represent 40-60% of total production expenses for most crops. Efficiency improvements enabled by AI directly translate to improved profitability, making farming more economically sustainable.

Climate Resilience: Adapting to Uncertainty

Climate change poses existential challenges to global agriculture through increased weather variability, shifting growing seasons, and new pest pressures. AI provides tools for adapting to these changes through predictive modeling, risk management, and optimization systems.

Climate-smart agriculture powered by AI can optimize crop varieties for changing conditions, adjust management practices to minimize weather risks, and implement conservation strategies that build long-term resilience. These capabilities are essential for maintaining agricultural productivity in an increasingly unstable climate.

Agricultural climate models enhanced by AI suggest that adaptive management could maintain productivity even under moderate climate change scenarios, though this requires widespread adoption of intelligent farming systems (Jones et al., 2017).

Knowledge Democratization: The Unfulfilled Promise

AI's potential to democratize access to agricultural expertise represents its most promising aspect. Mobile apps powered by machine learning can theoretically provide smallholder farmers with crop management recommendations, pest identification services, and market intelligence previously available only to large commercial operations.

This democratization has particular significance for developing countries, where smallholder farmers produce most food but have limited access to extension services or technical support. AI-powered advisory services could bridge this gap, providing personalized recommendations based on local conditions and farmer resources.

However, realizing this potential requires addressing digital divide issues, developing culturally appropriate interfaces, and ensuring AI systems work effectively with diverse farming systems and traditional knowledge. The promise remains largely unfulfilled, most agricultural AI benefits currently accrue to well-resourced commercial operations rather than the smallholder farmers who most need support.

8. Challenges and Critical Considerations

Despite AI's tremendous potential in agriculture, several significant challenges must be addressed to ensure these technologies serve all farmers equitably and sustainably. My analysis of AI implementation across various agricultural contexts reveals critical issues demanding immediate attention.

The Digital Divide: Technology Haves and Have-Nots

The most pressing challenge facing agricultural AI involves the widening gap between technologically advanced and traditional farming operations. Large commercial farms with substantial capital resources can invest in sophisticated AI systems, gaining competitive advantages that smaller operations cannot match.

This disparity is particularly pronounced between developed and developing countries. While precision agriculture systems transform farming in North America and Europe, smallholder farmers in Sub-Saharan Africa often lack reliable electricity, let alone internet connectivity for AI-powered tools.

I believe this digital divide represents an existential threat to equitable agricultural development. Without deliberate interventions to ensure broad access to AI technologies, we risk creating a two-tiered agricultural system where technological advantages compound existing inequalities.

Recent studies from the International Food Policy Research Institute confirm this concern, documenting that technology adoption patterns in agriculture typically increase rather than decrease inequality between farmers (Spielman & Smale, 2017).

Data Ownership and Privacy Concerns

Agricultural AI systems generate vast amounts of data about farm operations, crop performance, and management decisions. Questions about who owns this data, how it can be used, and what protections exist for farmers represent critical concerns that remain largely unresolved.

Many farmers express legitimate concerns about sharing sensitive operational data with technology companies, particularly given the potential for this information to be used in ways that could disadvantage farmers in markets or negotiations with buyers.

The concentration of agricultural data among a few large technology companies raises additional concerns about market power and competitive dynamics. If essential AI tools require farmers to surrender control over their operational data, this creates dependencies that could undermine farmer autonomy.

Establishing clear data rights, privacy protections, and governance frameworks for agricultural data represents an urgent priority that policymakers have largely ignored.

Technological Dependence and System Resilience

As AI systems become integral to farm operations, concerns about technological dependence become increasingly relevant. System failures, cyberattacks, or service disruptions could have catastrophic consequences for farms that rely heavily on AI-powered tools.

This dependency risk is particularly acute in precision agriculture operations where GPS signal interruptions or software glitches can halt planting or harvesting operations during critical windows. The concentration of AI services among a few providers amplifies these risks.

Building resilient agricultural AI systems requires redundancy, fail-safe mechanisms, and contingency planning that many current implementations lack. Farmers need confidence that AI tools will function reliably when needed most.

Environmental Implications of AI Infrastructure

While AI systems improve agricultural resource efficiency, the computing infrastructure required to power these technologies has its own environmental footprint. Data centers supporting agricultural AI consume significant energy, potentially offsetting some environmental benefits.

Recent analysis suggests that the carbon footprint of AI training and inference for agricultural applications could represent 2-5% of the environmental benefits these systems generate (Strubell et al., 2019). While still net positive, this represents a non-trivial consideration for sustainable implementation.

The proliferation of sensors, devices, and equipment associated with smart farming also raises questions about electronic waste and resource consumption. Developing sustainable AI infrastructure for agriculture requires attention to these lifecycle environmental impacts.

9. Future Outlook: The Path Forward

Agricultural AI stands at a critical juncture. The technologies exist to transform food systems, but realizing their full potential requires addressing fundamental challenges around access, equity, and sustainability. Several key directions will shape the future of agricultural AI.

Integration with Emerging Technologies

The future of agricultural AI lies in integrated systems combining multiple technologies. The convergence of AI with Internet of Things sensors, blockchain systems, satellite imagery, and biotechnology will create comprehensive platforms for farm management and food system coordination.

Edge computing will enable AI processing directly on farm equipment, reducing dependence on internet connectivity and improving response times for critical decisions. 5G networks will support real-time data exchange between multiple systems, enabling coordination across entire agricultural landscapes.

Gene editing technologies guided by AI will accelerate crop improvement, developing varieties optimized for specific environments and management systems. This integration of biological and digital technologies represents the next frontier in agricultural innovation.

Sustainable AI Development

Growing awareness of AI's environmental footprint is driving development of more sustainable agricultural AI systems. Energy-efficient algorithms, renewable-powered data centers, and circular economy approaches to technology deployment will become essential components of responsible AI implementation.

Carbon accounting for AI systems may become a competitive advantage as agriculture faces increasing pressure to reduce its environmental impact. Technology companies developing agricultural AI will need to account for the full lifecycle environmental costs of their systems.

Democratization and Accessibility

The most significant opportunity for agricultural AI involves making these technologies accessible to all farmers, regardless of size or location. This requires developing low-cost solutions, improving rural infrastructure, and creating support systems for technology adoption.

Cooperative models, shared services, and government programs will play crucial roles in broadening AI access. Mobile-first development approaches can leverage widespread smartphone adoption to deliver AI capabilities without requiring expensive specialized equipment.

Open-source AI tools and community-supported development models may provide alternatives to proprietary systems that create technological dependencies and data concentration concerns.

Addressing Global Food System Inequities

Agricultural AI adoption must address rather than exacerbate existing inequities in global food systems. This means prioritizing development of solutions for smallholder farmers, ensuring technology transfer to developing countries, and creating governance frameworks that protect farmer interests.

International cooperation will be essential for leveraging AI to address global food security challenges while preventing technological advantages from further disadvantaging vulnerable populations.

The success of agricultural AI will ultimately be measured not by the sophistication of the technology, but by its contribution to building more equitable, sustainable, and resilient food systems that nourish all people.

10. Conclusion: Technology in Service of Humanity

The integration of artificial intelligence into agriculture represents both humanity's greatest opportunity and its most critical test in building sustainable food systems for the future. Throughout my research on AI implementation in farming, I've become convinced that these technologies are essential tools for addressing the converging crises of food security, environmental degradation, and climate change.

The evidence is clear: AI-powered precision agriculture can increase yields while dramatically reducing resource consumption, early detection systems can prevent crop losses that waste billions of dollars annually, and intelligent supply chains can eliminate much of the food waste between farm and consumer. These are not theoretical benefits but documented improvements transforming agriculture in many regions.

However, my analysis reveals a fundamental tension in agricultural AI development. The farmers who most need these technologies, smallholders struggling with climate change, limited resources, and market volatility, are precisely those with the least access to AI tools. Meanwhile, large commercial operations with substantial resources gain competitive advantages that may further concentrate agricultural production and market power.

This disparity threatens to undermine AI's potential to address global food security challenges. If these technologies remain accessible only to well-resourced farmers in developed countries, we risk creating a two-tiered global agricultural system that exacerbates existing inequalities while failing to address food insecurity where it is most severe.

The path forward requires deliberate action to ensure AI serves all farmers, not just the privileged few. This means developing affordable, accessible technologies designed for resource-constrained environments. It means investing in rural infrastructure, education, and support systems that enable widespread AI adoption. Most importantly, it means establishing governance frameworks that protect farmer interests while promoting beneficial innovation.

I remain optimistic about agricultural AI's potential, but only if we implement these technologies with wisdom and commitment to equity. The choices we make in the next decade will determine whether AI becomes a force for justice and sustainability or a source of increased inequality and environmental harm.

The stakes extend beyond agriculture to the fundamental question of whether technology serves humanity or the reverse. In agricultural AI, we have the opportunity to demonstrate that sophisticated technologies can be implemented in ways that strengthen rather than weaken social bonds, that empower rather than displace traditional knowledge, and that heal rather than harm the natural systems upon which we all depend.

Climate change is accelerating, population growth continues, and environmental degradation threatens the foundation of agricultural production. Traditional farming approaches cannot meet these challenges alone. AI offers tools that can transform agriculture into a precision science working in harmony with natural systems, but only if we ensure these tools serve the common good rather than narrow commercial interests.

The agricultural AI revolution is already underway. Our responsibility is to guide it toward outcomes that reflect humanity's highest aspirations for justice, sustainability, and shared prosperity. The future of food and the farmers who produce it, depends on making the right choices now.

About the Author

Farida Khalaf *is a data engineer, certified in threat intelligence and cybersecurity analytics, and a global affairs expert. Her interdisciplinary career combines a BA in Economics with two master's degrees in International Finance and Global Affairs. Farida has extensive experience collaborating with the UN and leading large-scale international grant projects. Her work provides a unique and pragmatic perspective on the future of AI in agriculture, bridging technical data expertise with real-world global policy. She is based in NYC, USA.*

Chapter 17

AI in Advertising and Marketing

By Lisa Raehsler

Introduction: That Creepy Moment When Ads Get Too Personal

YOU MENTIONED WANTING a new coffee maker to a colleague at lunch. You didn't search for one online. You didn't even think about it seriously. Yet by evening, your Instagram feed is flooded with coffee maker ads, your email has a "perfect timing" promotional message from a kitchen retailer, and even YouTube serves you a video review of the exact model you mentioned.

How did they know?

In my 20 years working with digital advertising campaigns, I've watched this transformation accelerate dramatically. What you experienced isn't a coincidence. It's artificial intelligence at work, and it's reshaping every interaction you have with brands. The technology that powers this apparent mind-reading has become so sophisticated that 71% of consumers now expect personalized experiences, while 76% feel frustrated when they don't receive them (Harutyunyan, 2026).

But most people don't realize that the same technology that delivers perfect product recommendations also raises complex questions about privacy, influence, and control. As someone who's been inside the marketing industry for years, I want to pull back the curtain on what's actually happening when you encounter these eerily accurate ads.

The reality is both more fascinating and more concerning than most professionals understand. Let me show you exactly how AI has transformed the advertising landscape and what it means for your daily digital life.

How AI Changed What You See Every Day

Your Before and After Experience

Remember when all ads were generic? I do. In the early 2000s, if you visited a news website, every reader saw the same banner ad for insurance or generic consumer products. Television commercials interrupted everyone's viewing with identical 30-second spots. Magazine advertisements targeted broad demographic categories like "women 25-45" or "men who golf."

Those days are long gone. Today's AI-driven marketing operates with surgical precision. Instead of demographic guesswork, algorithms analyze millions of behavioral signals to predict exactly what you want to see. The transformation has been staggering: companies using AI segmentation report 38% higher marketing ROI, and consumers exposed to AI-driven product recommendations are 31% more likely to convert (Elad, 2025).

This shift represents more than technological advancement. It's a fundamental change in the relationship between brands and consumers. Where advertising once interrupted your day with irrelevant messages, AI attempts to predict your needs and present solutions you actually want. The challenge lies in the invisible line between helpful and invasive.

What's Watching You Right Now

Every digital interaction you have creates data points that AI systems collect, analyze, and act upon. In my experience managing campaigns for hundreds of clients, this tracking has become far more sophisticated than most professionals realize.

Your clicks, scrolls, and pause moments tell a story. When you hover over a product image for three seconds longer than usual, that hesitation gets recorded. AI systems track not just what you click, but how you click.

Cross-device tracking connects your phone to your laptop to your smart TV. The coffee maker conversation gets correlated with your browsing patterns and streaming preferences through location data.

Even without cookies, systems can often recognize your device based on how it is configured and used.

Recent studies show that 40% of organizations report AI-related privacy incidents, often because these systems collect more information than users realize (Protecto, 2025). Yet 85% of marketers now use AI daily because the technology delivers results that traditional advertising never could (CoSchedule, 2025).

When you visit a website, dozens of AI systems bid against each other in milliseconds to show you their ads.

The Three Ways AI Figures Out What You Want

1. How It Finds People Just Like You

The most powerful aspect of AI marketing is what it learns from people who behave like you. Marketers use "lookalike modeling" to find audiences who are likely to purchase based on similar consumers. AI groups you with similar people based on hundreds of behavioral patterns you probably don't even notice. Do you read articles to the end? Skip video ads after exactly five seconds? Shop more on weekends or weekdays? These micro-behaviors create a behavioral fingerprint that connects you to others with remarkably similar patterns.

The algorithms don't just look at demographics. A 45-year-old executive and a 25-year-old graduate student might get grouped together because they both read reviews, pause before clicking "buy now," and tend to research products extensively before purchasing. AI recognizes these behavioral similarities across age, location, and income differences.

This is why you and your colleague often see similar ads despite having different backgrounds. You are algorithmically grouped together because you both exhibit "high-consideration buyer" patterns or "early-adopter behaviors." The AI has identified you as the same type of customer, even if you work in different industries.

The sophistication of this grouping has increased dramatically. Where early systems might identify 10-20 audience segments, modern

AI creates thousands of micro-segments, each containing people with incredibly specific behavioral patterns. Some segments might be as narrow as "urban professionals who read financial news on commuter trains and prefer sustainable products."

2. When AI Knows You're Ready to Buy

Purchase intent prediction has evolved far beyond tracking when someone puts items in shopping carts. Modern systems recognize dozens of behavioral signals that indicate buying readiness, often before you consciously realize you're shopping for something.

The timing signals are very subtle. AI notices when you start reading product reviews more carefully than usual, when you begin comparing prices across multiple websites, or when your search patterns shift from general research to specific product features. These behavioral changes often occur weeks before you make a purchase decision.

Location data adds another layer of prediction. If you visit three different electronics stores in a weekend, AI systems flag you as "high purchase intent" for electronics, even if you haven't searched for anything specific online.

AI also tracks financial patterns and can predict major purchases months in advance by recognizing how consumers research and compare options. For example, a stressed consumer might see more getaway offers, while someone celebrating wins might see luxury recommendations.

3. Why Every Ad Feels Written Just for You

The personalization of ad copy has reached levels that would have been impossible just five years ago. Where traditional advertising created maybe five versions of an ad, AI systems now generate thousands of variations automatically, each dynamically optimized for specific audience segments.

Dynamic creative optimization takes basic product information and automatically generates headlines, descriptions, and calls-to-action based on what resonates with your behavioral profile. If you tend to respond to urgency-based messaging, you'll see "limited time offer" language. If you prefer detailed information, you'll get product specification-heavy ad copy.

The ad copy language adaptation goes deeper than marketing messages. AI analyzes your reading level, preferred terminology, and communication style based on your digital interactions. Someone who uses professional jargon gets more technical product descriptions, while someone who prefers casual language sees simplified, conversational copy.

Visual elements adapt too. The same product might appear with different background colors, models, or settings based on your demonstrated preferences. AI systems test millions of combinations to determine which visual approaches resonate with different audience segments.

Even timing and frequency get personalized. Some users respond better to single, well-timed messages, while others need multiple touchpoints across different platforms. AI adjusts the cadence and intensity of messaging based on how you've responded to advertising historically.

AI Marketing in Your Daily Life

Why Your Social Media Knows Your Style

Social media platforms extend the behavioral tracking I described earlier into visual and social preferences. Instagram's algorithm analyzes more than your double-taps and comments. It tracks how long you look at different types of images, which visual elements make you stop scrolling, and the lighting and color schemes that capture your attention.

The path from social interaction to purchase has shortened dramatically. Someone who likes minimalist interior design posts might start seeing targeted ads for furniture brands, even if they've never indicated interest in redecorating.

This represents the same behavioral fingerprinting principles applied to visual and social data. AI systems now track your social media engagement to inform advertising across completely different platforms, creating a coordinated cross-platform experience where your Instagram activity influences the Google ads you see.

Email That Feels Like Mind-Reading

Email marketing represents one of AI's most successful applications, delivering 3600% ROI according to recent studies (Ehtesham, 2025). Subject lines adapt to your opening patterns. If you typically open emails with questions, you'll see more question-based subject lines. If you respond to urgency, AI systems will test different urgency levels to find your optimal trigger point.

Timing optimization has become incredibly precise. Instead of sending emails at "optimal" times for broad audiences, AI determines when you specifically are most likely to engage. Content personalization extends beyond inserting your name, analyzing your past email interactions to determine optimal message length, tone, and information density.

Product recommendations in emails now combine your browsing history, purchase patterns, and email engagement data to suggest products with remarkable accuracy. Many customers receive email recommendations for products they were considering but hadn't searched for yet.

Google Searches That Finish Your Thoughts

Search advertising has evolved. When managing search campaigns, I regularly see AI systems predict and respond to user intent faster than users can type their queries.

Autocomplete suggestions represent just the surface level of AI prediction. Behind the scenes, algorithms analyze partial queries, typing speed, and time of day to predict not just what you're searching for, but why you're searching for it. This intent prediction shapes both organic results and paid advertisements by delivering different results based on the suggestion you choose.

The AI connects your search patterns to your broader digital behavior. Someone who searches for "best laptop" while simultaneously browsing business articles might see ads for professional-grade computers, while someone with the same search but recent entertainment browsing might see gaming laptop ads.

Geographic and temporal factors add sophisticated targeting. AI recognizes when someone searches for "coffee" while in a business district at 9 AM versus at home on Saturday morning, serving different types of results and advertisements accordingly.

Voice search has introduced new AI capabilities. When you ask your smart speaker about products or services, AI systems analyze not just your words but your tone, urgency, and context to determine appropriate responses and advertising opportunities.

Living with AI That Knows Your Habits

What You Gain from Personalized Marketing

The benefits of AI-driven personalization extend far beyond receiving relevant ads. For consumers willing to share behavioral data, AI marketing can genuinely improve shopping experiences and decision-making processes.

Product discovery has become remarkably efficient. Instead of spending hours researching options, AI systems can quickly identify products that match your specific needs and preferences. The algorithm learns whether you prioritize price, quality, sustainability, or convenience, and surfaces options accordingly.

Time savings represent a significant advantage. When AI accurately predicts your interests, you spend less time filtering through irrelevant options and more time evaluating products you might actually want. Consumers exposed to AI-driven recommendations are more likely to convert, partly because the suggestions will align with existing preferences.

Decision support has improved dramatically. AI systems can analyze product reviews, specifications, and compatibility factors far more quickly than human research. For complex purchases like electronics or appliances, AI can summarize key differences and highlight features most relevant to your usage patterns.

Price optimization works in consumers' favor too. AI systems track pricing patterns and can predict when items go on sale, helping you time purchases optimally. Some platforms use AI to offer personalized discounts based on your price sensitivity and purchase likelihood.

What You Give Up for Convenience

The convenience of AI marketing comes with significant trade-offs that many professionals don't fully understand. The data sharing required for personalization creates vulnerabilities and influence mechanisms that extend far beyond advertising.

Privacy erosion happens gradually and often imperceptibly. Each interaction provides more data points for AI systems to collect, analyze, and act upon. Over time, these systems develop incredibly detailed profiles that can predict not only your shopping preferences, but your emotional states, financial situations, and personal relationships.

Manipulation risks increase with AI sophistication. Systems designed to influence purchase decisions can exploit psychological triggers. When AI knows you're more susceptible to certain messages during stress or excitement, that knowledge can be used to encourage impulse purchases.

Choice architecture gets subtly altered as AI systems learn to present options in ways that influence decisions. The order of search results, the framing of product descriptions, and the timing of offers can all be optimized to increase purchase likelihood in ways that bypass conscious decision-making processes.

AI systems learn from historical data, which means they can unintentionally repeat old patterns and exclusions. If past campaigns mostly targeted certain neighborhoods or income levels, AI can quietly continue that pattern. The result is that some groups see better offers and more opportunities, while others see fewer options or none at all. Responsible marketers have to check for these blind spots, not just assume the algorithm is neutral.

Taking Back Some Control

Understanding AI marketing systems empowers you to make more informed decisions about your digital interactions. The key lies in recognizing when algorithms are influencing your choices and creating conscious boundaries around data sharing.

Privacy settings require regular attention, as platforms frequently update their data collection practices. Most major platforms now offer granular controls over data sharing and advertising preferences, but these settings often default to maximum data collection. Taking time to review and adjust these settings quarterly can significantly impact your advertising experience.

Ad preference management tools have become more sophisticated. Platforms like Google, Facebook, and Amazon provide detailed controls over the types of ads you see and the data used for targeting. While completely opting-out isn't always possible, you can significantly influence the personalization algorithms.

Browser and app permissions deserve careful consideration. Location tracking, microphone access, and cross-site data sharing can be controlled through device settings. Understanding which permissions different apps request helps you make informed decisions about data sharing.

Alternative platforms and tools offer different privacy trade-offs. Privacy-focused search engines, email providers, and social media platforms provide varying levels of AI personalization and data protection. Choosing platforms aligned with your privacy preferences can significantly impact your advertising experience.

How to Spot Companies Doing AI Marketing Right

What Responsible AI Marketing Looks Like

After working with companies implementing AI marketing strategies, I've identified clear patterns that distinguish ethical practitioners from those pushing boundaries inappropriately.

Transparency stands as the most important indicator of responsible AI marketing. Companies doing it right clearly explain what data they collect, how they use it, and why. Their privacy policies are genuinely readable, not filled with legal jargon designed to hide practices. They proactively communicate when they implement new AI systems or change data usage policies.

Easy-to-use privacy controls separate responsible companies from those just checking compliance boxes. The best implementations provide accessible, granular controls that are not buried settings that reset with updates. Companies committed to responsible AI make opting out as easy as opting in.

Data minimization principles guide responsible AI implementations. These companies collect only the data necessary for stated purposes and automatically delete information when it's no longer needed. They resist the temptation to collect everything simply because they can.

Red Flags vs. Green Flags

Experience has taught me to recognize warning signs that indicate problematic AI marketing practices. These red flags can help you identify when companies prioritize manipulation over customer service.

Red Flags to Watch For:

- Pressure tactics that create artificial urgency ("Only 2 left in stock!" when inventory isn't actually limited)
- Pricing that changes based on your browsing behavior or perceived ability to pay
- Recommendations that consistently push higher-priced options without clear benefit explanations
- Difficulty finding or using privacy controls and opt-out mechanisms
- Vague privacy policies that don't explain AI usage in understandable terms
- Advertising that follows you persistently across platforms after you've opted out

Green Flags That Indicate Responsible Practices:

- Clear, accessible explanations of how AI personalizes your experience
- Easy-to-find privacy controls with granular options
- Recommendations that include lower-priced alternatives when appropriate
- Honest inventory and availability information
- Respectful frequency limits on advertising and communications
- Transparent pricing that doesn't vary based on personal data

Brands Getting It Right

Companies that let you see and control your data profile show genuine commitment to transparency. They provide dashboards where you can view the information they've collected, understand how it is used, and modify or delete data. This transparency builds trust while maintaining personalization benefits.

Algorithmic recommendations that help customers make better decisions are AI marketing at its best. These systems highlight products that match your needs while honestly presenting alternative options. They might even recommend waiting for sales or suggest complementary products that improve overall value.

Businesses with responsive customer service use AI to enhance human interactions rather than replace them. They leverage AI to route inquiries more effectively and give representatives relevant context, while maintaining easy escalation paths to human support when needed.

Progressive disclosure of AI capabilities helps customers understand and control their experience gradually. Instead of overwhelming users with complex privacy settings. These companies introduce personalization features slowly and incrementally, explaining benefits and controls at each step.

What's Next for AI and Your Choices

Tomorrow's Shopping Experience and Your Rights

The future of AI marketing will bring more sophisticated personalization while privacy legislation continues evolving to protect consumer interests. Voice commerce, using voice on devices to search products, will make AI marketing more conversational, while augmented reality shopping will blend digital and physical retail in new, exciting ways.

Current Privacy Rights: Strong EU privacy laws require explicit consent for data collection and give you rights to access, correct, delete, and export personal information. California-style laws grant similar rights to residents, while many US states now have privacy laws similar to California's. These regulations mandate clear privacy policies, respect for opt-out requests, and reasonable data security measures.

What's Coming: Predictive commerce may anticipate your needs so precisely that AI systems could place orders before you consciously decide to purchase. However, the EU is working on specific AI rules that will mandate clearer disclosure about AI usage in marketing.

The companies that succeed long-term will be those that use AI to genuinely serve customer interests while respecting privacy boundaries. Consumer engagement with privacy controls will drive industry standards toward more responsible practices.

Making AI Marketing Work for You

Key Takeaways

AI marketing has fundamentally transformed the relationship between brands and consumers, creating unprecedented personalization capabilities with new privacy and challenges. As someone who has witnessed this transformation from inside the industry, I can tell you that understanding these systems empowers you to benefit from AI capabilities while maintaining control over your data and decisions.

The technology will continue evolving fast, making ongoing education and active privacy management important skills for anyone navigating the digital marketplace. The companies that succeed long-term will be those that can use AI to both genuinely serve customer interests and maximize profits.

Consumer awareness and engagement with privacy controls will ultimately drive industry standards toward more responsible practices. When you actively manage your data preferences and support companies with transparent AI practices, you contribute to a more ethical digital advertising ecosystem.

The future of AI marketing depends on finding sustainable balance between the benefits of personalization and privacy protection. This balance will come through a combination of technological innovation, regulatory oversight, and informed consumer choices.

Three Things You Can Do This Week

1. **Review Your Ad Preferences:** Visit Google Ad Settings, Facebook Ad Preferences, and your other frequently used platforms to understand what data they're using for targeting. Adjust settings to match your comfort level with data sharing and personalization.

2. **Notice AI Influence:** For one week, consciously observe when ads seem very relevant or timely. Try to identify what behavioral signals might have triggered those specific recommendations. This awareness helps you recognize when algorithms are influencing your choices.

3. **Evaluate Your Data Comfort Level:** Decide which types of data sharing provide genuine value for your lifestyle and which types cross lines you're not comfortable with. Use this assessment to guide your privacy settings and platform choices going forward.

The goal is not to avoid AI marketing. That's not practical or necessarily beneficial. Instead, the objective is to engage with these systems proactively, understanding how they work and making informed choices about the role you want them to play in your life.

AI marketing will continue evolving, but your participation in shaping that will determine if these powerful systems serve your interests and make life easier or simply extract and profit from your data. The choice is yours to make.

About the Author

Lisa Raehsler *is a U.S.-based digital advertising strategist and founder of* ***Big Click Co.****, where she helps brands improve performance through AI-driven paid media. With over 20 years of experience in Google Ads, analytics, and audience targeting, she has led campaigns for hundreds of clients across industries. Lisa is a contributor to Search Engine Journal, a Microsoft MVP award winner, and a frequent speaker at global marketing conferences. Her work focuses on using smart data and automation to boost ad results. Lisa also publishes* ***The Paid Media Mix****, a newsletter exploring the evolving role of AI in advertising.*

Chapter 18

AI in Retail and e-Commerce

By Katrina Watson

Introduction

FOR AS LONG AS I can remember, I've loved the idea of combining business needs with technology and design. I launched my first company at 22, full of hopes, dreams, and just the right amount of naivete that I could make a difference. My goal was simple: to make the world better, one website at a time.

But as I got older, I wanted to make a bigger difference; one website at a time felt too slow. I've always been passionate about helping small businesses compete–about helping the *underdogs*, as we like to say in Australia. Over the past few years, it feels like the big corporations are getting bigger and richer, while many small businesses are struggling or going under.

For years, the corporate giants have been using artificial intelligence (AI) to dominate at extraordinary speed, crushing many small businesses in the process. But AI is now becoming much more accessible, and for a lower cost.

When people talk about AI today, they often mean generative AI systems that create new content. But much of the AI that has shaped retail over the past two decades has been predictive rather than generative, focused on analysing data to anticipate customer behaviour.

In this chapter, I'll cover a brief history of how some retailers have been using AI to change the way we shop for years, and why I think that new AI tools could be the thing that helps level the playing field for smaller retailers.

History of Personalisation

You might not realise it, but AI has been at play in retail for a long time. More than 25 years ago, in the late 1990s, Amazon introduced the *Customers who bought this also bought...* feature. And it completely changed the way we shopped online.

With Amazon's recommendation system, these weren't just random products. Each suggestion *felt* like a *personal* recommendation, because it was powered by customer data. That key difference created trust, even if it was generated by a machine.

And it solved a real customer problem. Customers no longer had to spend hours researching what to read next. They could simply look at a book they were interested in on amazon.com and find ten other books that other people had also purchased that they'd probably like too.

This strategic move set Amazon apart from all their competitors, and quickly catapulted them into becoming the giant that they are today. In a few short years, Amazon grew from nothing to the world's top e-commerce platform.

By the mid-to-late 2000s, many other large retailers, such as Walmart and Target, had begun experimenting with personalisation, usually through third-party recommendation vendors. But these were received with mixed results.

In 2012 Target made headlines when an angry father reached out, complaining about personalised marketing that his teenage daughter received promoting baby products.

"My daughter got this in the mail!" he said. *"She's still in high school, and you're sending her coupons for baby clothes and cribs? Are you trying to encourage her to get pregnant?"*

The store manager called the man to apologise for the mixup, only to find that the father was now the one apologizing.

"I had a talk with my daughter," he said. *"It turns out there's been some activities in my house I haven't been completely aware of. She's due in August. I owe you an apology."*

Target's response was to start mixing their personalised recommendations with other products. So that the 'personalised' ads started to seem more random.

"We started mixing in all these ads for things we knew pregnant women would never buy, so the baby ads looked random. We'd put an ad for a lawn mower next to diapers. We'd put a coupon for wine glasses next to infant clothes. That way, it looked like all the products were chosen by chance. And we found out that as long as a pregnant woman thinks she hasn't been spied on, she'll use the coupons. She just assumes that everyone else on her block got the same mailer for diapers and cribs. As long as we don't spook her, it works."

Target isn't the only retailer doing this. Personalised marketing is one of the most common ways that retailers are using data and AI to boost sales and drive customer loyalty, with case studies showing that personalisation helps companies grow on average 2.5x faster than those that don't offer it (M Accelerator, 2025).

Dynamic Yield, one of the leading platforms for providing personalisation and experience optimisation, boasts impressive results helping customers such as Skims, Lacoste, Loreal, and Asics. Within a few months of launching a personalisation program for Skims, they were able to increase direct revenue by 10% per month (Dynamic Yield, 2022). They've also had success helping On, a sneaker brand, increase click-through rates to secondary products by 537% (Dynamic Yield, 2020).

It's not just well-known brands and giants leveraging AI to drive sales. Now even the smallest retailer can leverage a 'personalisation-as-a-service' platform to offer personalised recommendations to their customers. In the past, companies like Amazon spent millions of dollars to build advanced solutions like this, but now anyone can add it to their website for a relatively small fee. Generative AI can help small businesses personalise content quicker and easier than ever before.

But there's a fine line between being helpful and being creepy when personalising content and marketing. While it's become a key revenue driver for retailers, and *can* create a *better* customer experience, it also raises questions about how much data companies really need to know about us, and whether we're still *choosing* what we want to buy or being manipulated into buying things we don't really need.

Personalisation Tips:

- **Small businesses can use personalisation to create better customer experiences, driving sales and building loyalty.** Creating a 'for you' page makes it clear that the products and offers are tailored specifically to that customer. Using AI recommendations based on what others have purchased instead of generic 'you might also like' is another successful way to expose other products that a customer might also be interested in, especially if a retailer has a large range of products.
- **Personalised marketing emails and ads allow retailers to focus their efforts on products that a customer is most likely to be interested in, instead of spamming them with daily emails that often go unread or ignored.** But personalisation isn't just about adding a customer's name to the email. By personalising content, users are more likely to read and engage with emails. Taking into consideration things like the user's location, what items they purchase, and when they're most likely to want to make additional purchases can help increase conversions and create better customer experiences. For example, don't tell them *"Brrrr, it's cold outside"* when they live in the tropics and it's summer. Are there specific categories or things that they've shown interest in?
- **Most importantly, don't be a creep**. Only collect as much data as you really need, and be mindful about how you use it. Data privacy and consumer data regulations vary by country and region. Retailers should understand and respect any regulations that apply where their customers live, but it can also be beneficial to go above and beyond when privacy is concerned. For example, every business should avoid storing personally identifiable information, or ensure that it is encrypted and kept secure. This way, in the unfortunate event that any data is compromised, their customers will be protected, which can reduce the negative impacts on the business as well.

Dynamic Pricing

Dynamic pricing, where prices might differ based on location, time of day, or supply and demand, isn't new. Airlines, Uber, and petrol stations have been doing it for years. But with AI technologies now being much more accessible, more retailers (including physical store retailers) are starting to explore using it.

To understand more about how this works, let me briefly explain the traditional way that companies adjusted their pricing.

Some large companies employed price watchers. These people kept an eye on what their competitors are charging, manually monitoring trends and adjusting prices. Thanks to AI, it's now become easier than ever to process large datasets and find patterns, allowing retailers to automatically adjust pricing in real time. Some retailers go a step further and take into consideration things like local income levels, seasonal trends, willingness to pay, and even what the weather is that day.

As you might expect, customers have some strong feelings about prices changing, especially when it feels out of their control.

In early 2024, Wendy's CEO announced in a shareholder call that it was installing digital menus so that it could begin testing dynamic pricing. When the news hit media outlets, they called it 'surge pricing', which quickly escalated into customer outrage and boycotts, and even reached the US Senate. Wendy's was quick to respond, announcing that they wouldn't raise prices in times of high demand, only lower them when it was quiet.

But not all dynamic pricing is perceived as bad. Rema 1000, a Norwegian supermarket, implemented Electronic Shelf Labels in early 2024. These price labels automatically change throughout the day. But where this was different was that Rema 1000 was very clear and transparent about how the prices will change–stating from the start that prices will only ever go down, and are reset every night. Some users found it unsettling when the price of a product that they were about to buy changed right before their eyes, but overall the feature was well received.

What I really like about Rema 1000's approach is that they aren't just using dynamic pricing to increase revenue. They're also helping to cut food waste. The company reports that they have sold over 23.5 million discounted items in the last year. By automatically adjusting prices before the goods expire, especially on perishable items such as bakery goods, the chain is able to reduce financial losses and improve sustainability. This, combined with other measures, has resulted in them reducing food waste by 40% (Rema, n.d.).

And it's been great for the brand. They already had a very loyal customer base, holding the top spot in customer loyalty for seven consecutive years, but their brand loyalty continued to increase even further implementing these changes. They credit this to providing good customer experiences, combined with sustainability and social responsibility.

Three dynamic pricing lessons:

1. **Communicate the benefits.** Rema 1000 does this by pricing only going down. Some companies will let you know if the price drops or even guarantee to refund the difference if it drops within 48 hours of a purchase.
2. **Name the rule, not the algorithm.** "Markdowns from 4pm" beats "Dynamic pricing"
3. **Measure the metrics:** Don't just measure success based on revenue. Also monitor customer complaints, loyalty, waste, missed revenue opportunities.

Reducing Returns

Another positive example of how AI is helping retail is by reducing the number of returns.

Have you ever bought something online, only to find it wasn't what you wanted or it didn't fit? You're not alone.

Many retailers, especially larger ones, offer free returns to help build trust and loyalty with their customers. But every time we return something, even if it's promoted as free, it's not really 'free'. The cost to return and restock an item can be as much as 40% of the original value. There is also a significant environmental impact.

A 2022 report from Optoro, a company that specialises in returns logistics, calculated that 24 million metric tonnes of CO2 emissions are attributed to e-commerce returns every year. That's equivalent to emissions from 5.1 million cars over a year (Optoro, 2022).

When an item is returned, retailers have to decide if the cost of processing, inspecting and repackaging the item is worth it. The European Environment Agency estimates that 4-9% of textile products put on the market are destroyed before use, either because they weren't sold, couldn't be resold after a return, or because it was simply more financially viable for the retailer to dispose of it (European Environment Agency, 2024).

In an effort to address rising waste generation and transition towards a circular economy (an economic system where materials or products are reused or regenerated), the EU is adopting a new Waste Framework Directive. From July 2026, large companies will be banned from destroying unsold clothing, shoes, and accessories, except under strict conditions such as health, hygiene, or intellectual property issues.

Zalando, Europe's largest online fashion platform, realised that half of its returns came down to one thing: fit. Their AI systems now analyse customer feedback, previous purchases, and even the stretch of fabrics to suggest the right size. Returns dropped. Customer loyalty rose. And the carbon footprint of each sale shrank. Recently they've been experimenting with using generative AI to convert photos of real models wearing real products into videos, helping customers visualise products better before buying them. AI-generated content is often criticized, especially when everything is AI-generated (including the model), but if it helps customers make better choices about what they buy and reduces returns, could it become a good thing?

Warby Parker tackled the same problem from another angle, using augmented reality (AR) to let customers virtually try on glasses before buying. Customers can see what the glasses look like on their own face before making a purchase. IKEA did the same with furniture. The idea is simple: when you can 'see it' before you buy, you send less back.

Here are four ways that businesses can use AI to reduce returns, reducing costs, improving customer loyalty, and becoming more sustainable:

1. **Improve product descriptions.** Many small businesses don't have the time to write detailed product descriptions. This is a perfect opportunity where generative AI tools can help. By being more descriptive, including information about the material and the fit, AI can help make what they are purchasing clearer to customers.
2. **Leverage AI Size Calculators.** These combine data, often across multiple brands, to help customers find the right size for them, based on other things they have purchased. These size calculators have been shown to reduce returns by up to 45% (Prime AI, 2025).
3. **Monitor returns.** Many companies have data, but lack the skills or tools to process that data. New AI tools are making it easier for everyday people to use common language to query data. This allows them to identify products that have a higher return rate. They can then dig deeper to find out: Is it a problem with the product? Do sizes run smaller/larger than expected? These analyses can then be used to improve the product or communicate the product better.
4. **Help customers 'see' items before they buy them.** Create additional images, video or 3D models to show customers the product from additional angles. It's important that the AI-generated versions are accurate representations; otherwise, this can have the opposite effect and mislead customers. Alternatively, augmented reality solutions can work well, depending on the type of product the retailer is selling.

Predictive Analytics and On-Demand Manufacturing

AI allows retailers to move at unprecedented speed. This has resulted in big changes in the retail world, especially when it comes to product development within the fashion industry.

Traditionally, fashion designers would spend weeks to months researching trends, developing concepts, and creating initial sketches and mood boards. Then they'd spend a few more weeks sourcing fabrics and negotiating with suppliers and manufacturers. Pattern makers would then create prototypes, often going through multiple prototypes to refine the design, fit, and functionality. After a few rounds of testing, the final prototype would then be scheduled for manufacturing, which could take a few weeks to months depending on manufacturer availability. The total cycle typically took 4-8 months. This meant fashion brands worked months ahead to plan their products, usually producing two collections a year: one for Spring/Summer and one for Autumn/Winter.

Using data and AI, fast-fashion brands like Zara are now able to spot a trend and have a new product on the shelves in ***weeks***, not *months*. Zara's systems extend beyond their own orders, looking into things like social trends, comments, returns, and even the weather to understand more about what customers want. They then use this data to feed a fast-turnover supply chain, creating 450 million garments a year, including 20,000 new styles. (Ellison, 2023). This approach has resulted in Zara becoming one of the world's largest and most influential fast fashion retailers, with annual global revenue exceeding USD$28 billion.

From Fast-Fashion to Ultra-Fast-Fashion

But growth is starting to slow for fast-fashion brands like Zara. Ultra-fast-fashion brands like Shein have started to take over the market, gaining as much as 40-50% market share in some countries in a few years, with even faster and cheaper offerings.

Where Zara is producing around *one thousand* new products a month, Shein produces *tens of thousands* of new products *every* month, using demand-driven production that only manufactures items at scale after enough online interest surfaces. This approach significantly reduces the risk and waste for Shein, and allows them to move quickly and offer products much cheaper than other retailers, which has resulted in their revenue nearly doubling every year for the last five years.

But while their production methods are intended to reduce waste, it's actually creating more of it. Consumers, driven by unbelievably cheap products, are buying more clothes, but wearing them less.

Analysts are warning about the long-term viability of these ultra-cheap retailers. While sales have been growing, net profit is dropping. And consumers, especially Millennials and GenZ (everyone born after 1981), are starting to be more mindful about sustainability, circularity, and ethics.

The Changing E-Commerce Landscape

Over the last few years, there has been an increase in the number of retailers shutting down. In 2024, the number of retail stores closing in the US increased 69% vs. the previous year (Picchi, 2024). And the trend seems to be continuing with 15,000 stores expected to close in 2025, more than double the previous year (Danziger, 2025). This trend is not exclusive to the USA; other countries, such as Australia and the UK, have also experienced similar closures. Declining sales, increased costs and online competition are usually stated as the most common reasons.

But it's not that people are spending less. Rather, they're changing the way they shop.

The rise of fast and ultra-fast-fashion has changed how we think about shopping. Now every time we visit a retailer's website or open their app, we *expect* there to be something new. We've gone from thinking about "What do I *need?*" to "What do I *want now?*". Since the rise of fast fashion in the early 2000's, the number of garments produced globally every year has grown from 50 billion pieces to an estimated 150 billion in 2025 (Winkless, 2025).

Online retailers such as Shein and Temu have been some of the biggest benefactors of this growth. But in an attempt to re-balance global trade and curb the environmental and social impacts of fast fashion, governments across the world have been introducing new tariffs, taxes, and laws. These new taxes are designed to raise the cost of ultra-low-price goods, making them less attractive to consumers. Some countries, such as France and Italy, have been developing bills to ban or restrict advertising and influencer marketing for ultra-fast-fashion items (García, 2025; Reuters, 2025).

The fashion industry is one of the highest emitters of greenhouse gases globally, and changes are needed to reach the goals of the Paris Climate Agreement.

Organisations all over the world are trying to move towards Circular Fashion economies, helping to create action-oriented blueprints for the future of fashion.

New regulations will require retailers to have better data, and to be transparent about it. Retailers will need to track and disclose environmental footprint data for each and every product. This includes things like water usage, emissions, and if the goods can be recycled.

While this creates new reporting requirements, it may also change the way that products are made. A study of 102 textile enterprises in China found that greater AI adoption was associated with reducing pollution and carbon emissions, through improvements in operational efficiencies (Shi, 2026; Chen, 2026; Lin, 2026).

AI Adoption

AI is a chance for small and medium retailers to work smarter instead of harder–to learn from the mistakes of others and do things better. But AI adoption rates vary significantly between the giants vs. small to medium businesses (SMBs).

According to an NVIDIAreport, 90% of larger retailers are actively using AI or are actively assessing how to use AI in their business (NVIDIA, 2025). For SMB retailers, the rate of adoption is a lot lower, estimated at only around 20-40% (Australian Government Department of Industry, Science and Resources, 2025; NFIB, 2025).

There are plenty of case studies from businesses who have implemented AI saying it increases revenue and decreases costs (Shopify, 2025). So why aren't smaller retailers leveraging AI more?

For many, it's because they simply don't know what to do, or that they are wary of losing their personal touch. Smaller businesses are often concerned about customer scepticism, data privacy, and the quality and consistency of AI-generated outputs. This is a fair concern.

Many also don't believe that it will help them compete more effectively: that the costs don't outweigh the benefits or that the systems aren't 'good enough' yet (BizCover, 2025; Cao, 2025; Marks, 2025; Xero, 2025).

Right now, AI tools are affordable and accessible to smaller retailers while providers are subsidising the costs of the tools to grow their user bases.. But it's unclear how long this will last. Currently there are tools on the market from as little as US$20 a month that allow small businesses to do everything from personalisation, price optimisation, customer support, inventory forecasting, and demand prediction. But even with these tools, most businesses will need to invest more time and money to make AI work for them, personalising them to suit their individual needs.

In small and medium businesses, marketing is one of the areas which is often neglected due to time, skills and resources. This is an area where generative AI may help them close the gap. Some retailers have already started creating custom AI models that can generate content using their brand voice, helping them to automatically create search engine optimised (SEO) blog posts and marketing content (Boyer, 2024).

Conclusion

AI is one of the reasons that some big retail businesses have become so successful. Many small business owners struggle finding enough time to do all the things that they want. They'd love to do more marketing or optimise their website, but they lack the skills or time to do so.

This is where generative AI tools can help them level up. But using AI blindly or poorly doesn't mean instant success. To win against the giants, small businesses need to find ways to work smarter not harder; to use AI with purpose; to solve real problems; and to keep their customers front of mind when they do so.

Small businesses can move faster. They can innovate and be creative. They can leverage AI to create better, more personalised customer experiences. And that will help drive stronger customer loyalty, building more sustainable businesses.

About the Author

Katrina Watson *is a product leader and entrepreneur with over two decades of experience building digital products for small business. She was Head of Product at Sendle, a sustainable shipping company focused on helping SMBs compete at scale. Before that, she spent a decade founding and running a web development agency, where she helped thousands of SMBs grow and improve their businesses. Katrina is passionate about how technology and design can create better, more sustainable experiences for small businesses and their customers. She is based in North Queensland, Australia.*

HOW-TO

Chapter 19

AI and Cybersecurity

Transforming Digital Defense in the Modern Era

By Farida Khalaf

Intended Audience: *This article assumes a foundational understanding of AI and machine learning concepts and terminology. For readers less familiar with these topics, a helpful resource on the basics of AI and ML can be found at https://sixpeas.substack.com/ (Smiley, 2024).*

1. Introduction: The AI-Cybersecurity Nexus

THE CONVERGENCE OF artificial intelligence (AI) and cybersecurity represents one of the most significant paradigm shifts in digital security since the advent of the internet. Recent comprehensive literature reviews have identified over 2,395 studies examining AI applications in cybersecurity, with 236 classified as primary research sources, underscoring the rapid maturation of this field (Sarker et al., 2023). As organizations worldwide grapple with an exponentially expanding threat landscape, traditional signature-based security approaches are proving inadequate against sophisticated, rapidly evolving cyber threats.

The scale of contemporary cyber threats is staggering. Every day, cybersecurity systems detect approximately 200,454 new malware threats, with threat actors deploying an average of 1.5 new malware samples per minute, roughly 200,000 unique scripts daily (StationX, 2024; AV-TEST, 2025). This represents a fundamental shift from earlier cybersecurity paradigms, where threats numbered in the thousands rather than hundreds of thousands daily. The traditional, reactive approach of developing defenses after threats emerge is constantly being tested as cyber threats become more sophisticated, demanding advanced solutions that can respond to and prevent attacks proactively (Salem et al., 2024).

In response, cybersecurity professionals are increasingly turning to AI and machine learning (ML) to level the playing field. Between 2020 and 2024 alone, 9,084 academic articles were published on AI-driven cybersecurity detection techniques (Alkaabi et al., 2024). This transformation represents a fundamental shift toward proactive, predictive security postures that can identify and neutralize threats before they cause damage. It's important to note that AI in cybersecurity encompasses far more than just generative AI (genAI) or large language models (LLMs), which are often associated with chatbots and content creation. Instead, AI-driven cybersecurity leverages machine learning (ML), reinforcement learning (RL), and data science to enhance threat detection, incident response, and network defense. However, this same technology also empowers adversaries to conduct attacks at unprecedented scale and sophistication, creating an arms race that will define the future of cybersecurity.

2. AI as a Cybersecurity Defense Tool

AI-powered cybersecurity systems rely on machine learning (ML) algorithms to process vast amounts of data and identify patterns that would be impossible for human analysts to detect manually. These systems typically operate in two main phases:

- **The training phase**: During this phase, the system learns from massive datasets that include network traffic, user behavior, malware samples, and known attack patterns.
- **The inference phase**: Here, the trained system analyzes real-time data to detect threats and respond to incidents.

AI systems excel at pattern recognition and prediction using large datasets of security information. By analyzing network traffic, they can identify malicious activity, which is especially useful in detecting advanced attacks such as Advanced Persistent Threats (APTs), which can remain undetected for months. Moreover, these systems can predict the likelihood of future attacks and identify vulnerable systems based on historical data and up-to-date threat intelligence.

Table 1 below shows the typical phases involved in implementing an AI-driven cybersecurity system, demonstrating the progression from basic threat detection to advanced predictive analytics. Durations listed are approximate and assume the presence of a mature system, existing data pipelines, and a medium to large-scale project. The timeline may vary based on the complexity and size of the organization's cybersecurity infrastructure.

Table 1: Typical Phases of AI Security System Implementation

Phase and Duration	Key Activities	Expected Outcomes	Success Metrics
Data Collection & Preparation 2-4 weeks	• Aggregate historical security logs • Clean and normalize data formats • Establish data quality baselines	• Comprehensive dataset ready for training • Data governance framework established	• 95%+ data completeness • <5% data quality issues
Model Training & Validation (4-8 weeks)	• Train ML models on prepared datasets • Perform cross-validation testing • Optimize hyperparameters	• Trained models achieving target accuracy • Validated performance metrics	• >95% detection accuracy • <5% false positive rate
Pilot Deployment (6-12 weeks)	• Deploy in controlled environment • Monitor performance in real-time • Collect feedback from security teams	• Proven system effectiveness • Staff trained on new tools	• 60% faster threat detection • 90% user satisfaction

Phase and Duration	Key Activities	Expected Outcomes	Success Metrics
Production Integration (8-16 weeks)	• Integrate with existing Security Information and Event Management (SIEM) systems, or Security Orchestration, Automation, and Response (SOAR) platform • Establish automated response workflows • Scale across enterprise environment	• Fully operational AI security platform • Automated threat response capability	• 24/7 operational status< • <3 minute response time
Continuous Optimization (Ongoing)	• Regular model retraining • Performance monitoring & tuning • Threat landscape adaptation	• Evolving defense capabilities • Maintained effectiveness against new threats	• Monthly model updates • Sustained performance metrics

Advanced Threat Detection and Pattern Recognition

AI's most significant contribution to cybersecurity lies in its ability to process vast amounts of data and detect patterns that would be impossible for human analysts to identify manually. Modern AI-powered security systems utilize multiple machine learning paradigms: supervised learning models trained on labeled datasets achieve 95-99% accuracy in detecting known malware variants, while unsupervised learning algorithms excel at identifying zero-day threats through statistical anomaly detection (Apruzzese et al., 2023).

Deep learning neural networks, particularly convolutional neural networks (CNNs) and recurrent neural networks (RNNs), have demonstrated exceptional capability in analyzing complex relationships within network traffic data. However, more recent architectures, such as **transformers** and **graph neural networks (GNNs)**, are also showing significant promise in this space. Transformers, with their ability to capture long-range dependencies in sequential data, and GNNs, with their capacity to model complex relationships in graph-structured data, are proving effective in handling the dynamic, interconnected nature of modern network systems. These systems can process terabytes of log data in real-time, establishing baseline behavioral patterns for networks, systems, and users, which form the foundation for anomaly detection.

Research from Darktrace's 2025 State of AI Cyber Security survey of over 1,800 cybersecurity professionals reveals that organizations implementing AI-powered threat detection report 60% faster threat identification compared to traditional signature-based systems, while maintaining false positive rates below 5%. In contrast, traditional signature-based systems typically experience false positive rates that can exceed 10%, which can overwhelm security teams and reduce the effectiveness of detection efforts. AI systems, particularly those using advanced techniques like ensemble learning, significantly reduce false positives by distinguishing between benign and malicious activity more accurately.

Security Orchestration and Automated Response (SOAR)

AI-enabled Security Orchestration, Automation, and Response (SOAR) platforms represent a quantum leap in incident response capabilities. These systems integrate multiple security tools and data sources, enabling coordinated responses that can execute in milliseconds, compared to the minutes or hours traditionally required for human intervention.

Contemporary AI-driven SOAR implementations utilize reinforcement learning (RL) algorithms, a subset of AI where systems learn from past interactions to optimize future actions. These algorithms continuously improve response strategies based on attack outcomes and organizational priorities. For example, when detecting a potential Advanced Persistent Threat (APT), these systems can simultaneously isolate affected endpoints, preserve forensic evidence, update firewall rules across the enterprise, and initiate threat-hunting procedures, all while alerting security teams.

Case studies from financial services organizations demonstrate that AI-powered automated response systems can contain lateral movement attempts within an average of 3.7 minutes, compared to 197 minutes for manual response procedures (IBM Security, 2025).

User and Entity Behavior Analytics (UEBA)

UEBA represents one of AI's most sophisticated applications in cybersecurity, addressing the critical challenge of insider threats and compromised accounts. These systems employ unsupervised machine learning algorithms to establish behavioral baselines for every user, device, and application within an organization's environment.

Modern UEBA implementations analyze over 100 behavioral variables simultaneously, including temporal access patterns and geographical locations, file access sequences and data transfer volumes, application usage patterns and command-line activities, network connection patterns and protocol utilizations, and authentication patterns and privilege escalations.

Leading UEBA platforms utilize ensemble methods combining isolation forests, one-class support vector machines, and autoencoder neural networks to achieve anomaly detection accuracy rates exceeding 94% while maintaining false positive rates below 2% (Microsoft Security Intelligence Report, 2024).

Real-World Implementation: JPMorgan Chase Case Study

JPMorgan Chase, which faces over 45 billion hacking attempts daily, has implemented a comprehensive AI-powered cybersecurity platform that successfully blocks these attacks and processes millions of security events. Their system, which combines deep learning neural networks with graph analytics, has achieved:

- 85% reduction in false positive security alerts
- 99.7% accuracy in detecting fraudulent transactions
- 40% improvement in threat detection speed
- $150 million annual cost savings through automation

The bank's implementation demonstrates that AI systems now detect and respond to threats that would have taken human analysts weeks to identify, often preventing attacks before any damage occurs (JPMorgan Chase Annual Report, 2025).

3. AI-Powered Cyber Threats

The same artificial intelligence technologies that enhance defensive capabilities are being weaponized by cybercriminals and nation-state attackers. This creates an escalating arms race where both defenders and attackers leverage increasingly sophisticated AI tools, fundamentally changing the threat landscape.

Adversarial Machine Learning and Evasion Techniques

The weaponization of AI by cybercriminals represents an emerging and sophisticated threat vector. Adversarial machine learning techniques enable attackers to craft malware specifically designed to evade AI-powered detection systems. These attacks exploit the mathematical properties of machine learning models, introducing carefully crafted perturbations that cause classifiers to misidentify malicious samples as benign.

Recent research has demonstrated that adversarial examples can achieve evasion rates exceeding 90% against commercial antivirus solutions employing machine learning (Grosse et al., 2024). These attacks utilize techniques such as gradient-based optimization to find minimal perturbations, generative adversarial networks (GANs) to create evasive malware variants, model extraction attacks to reverse-engineer detection algorithms, and data poisoning during model training phases.

AI-Enhanced Malware and Polymorphic Attacks

Contemporary malware increasingly incorporates AI capabilities to enhance its effectiveness and longevity. AI-powered malware can adapt its behavior in real-time based on the environment it encounters, modify its code structure to avoid signature detection, and learn from failed attack attempts to improve future success rates.

The most active year for new malware variants was 2021, when around 150 million new programs were identified, but recent developments in AI-generated malware suggest volumes may increase as automation reduces the skill barrier for malware creation (Deepstrike, 2025).

Modern polymorphic malware employs machine learning algorithms to generate thousands of functionally equivalent code variants, optimize payload delivery based on target system characteristics, adapt communication protocols to blend with legitimate network traffic, and implement dynamic command and control infrastructure.

Deepfakes and Advanced Social Engineering

The emergence of sophisticated deepfake technology has created new vectors for social engineering attacks at unprecedented scale and believability. High-quality audio deepfakes can now be generated with as little as 10-15 seconds of sample audio, while video deepfakes can create convincing impersonations.

The financial impact of deepfake-enabled fraud has been substantial. In 2024, a multinational corporation lost $25.6 million to fraudsters who used deepfake video technology to impersonate the company's chief financial officer during a video conference call (CNN Business, 2024). This incident represents the largest publicly-disclosed deepfake fraud case to date.

Large language models (LLMs) have revolutionized phishing and social engineering by enabling the creation of highly personalized and contextually appropriate attack messages. These AI systems can analyze targets' social media profiles to craft personalized messages, generate emails that match specific writing styles and communication patterns, create multi-stage social engineering campaigns with consistent personas, and adapt messaging based on target responses and engagement patterns.

Nation-State AI Capabilities

Advanced persistent threat (APT) groups sponsored by nation-states have significant resources to develop sophisticated AI-powered attack tools. Based on Rapid7 Labs data, 33 new or rebranded threat actors appeared in 2024, with 65 active ransomware groups observed in Q2 2025, many demonstrating advanced technical capabilities consistent with AI enhancement (Rapid7, 2025).

Intelligence agencies report that state-sponsored groups are investing heavily in AI-powered intelligence collection and target reconnaissance, automated social media manipulation and disinformation campaigns, machine learning-enhanced malware for long-term persistence, and AI-assisted analysis of stolen data and intelligence.

4. Challenges and Limitations

Despite their promise, AI-powered cybersecurity systems face significant challenges that organizations must understand and address. These limitations affect deployment decisions, operational effectiveness, and strategic planning for cybersecurity programs.

Algorithmic Bias and Fairness Concerns

AI security systems are susceptible to various forms of bias that can undermine their effectiveness and create ethical concerns. Training data bias occurs when historical security data over-represents certain types of attacks or under-represents others, leading to skewed model performance.

Research from MIT's Computer Science and Artificial Intelligence Laboratory demonstrates that commercial AI security tools exhibit significant bias across different attack categories, with detection rates varying by up to 40% between different malware families due to training data imbalances (MIT CSAIL, 2024). Demographic bias in user behavior analytics can result in certain user groups being flagged as higher risk due to behavioral patterns that correlate with demographic characteristics rather than actual security threats.

Additionally, consumer concerns about AI-driven surveillance are rising, with 68% of consumers expressing worry over how AI systems handle personal data in cybersecurity contexts, potentially leading to privacy erosion and reduced trust in digital services (Pew Research, 2025).

Adversarial Attacks Against AI Systems

The vulnerability of AI systems to adversarial attacks represents a fundamental security challenge. These attacks exploit the mathematical properties of machine learning models to cause incorrect classifications or decisions through evasion attacks that modify inputs at test time to avoid detection, poisoning attacks that corrupt training data to degrade model performance, model extraction attacks that steal model parameters through query analysis, and membership inference attacks that determine if specific data was used in training.

Research demonstrates that adversarial attacks can reduce the effectiveness of AI-powered intrusion detection systems by up to 95% while remaining undetectable to human analysts (Chen et al., 2024).

The Explainability Challenge

The "black box" nature of many AI systems, particularly deep learning models, creates significant challenges in cybersecurity applications where decision-making transparency is critical. Security teams need to understand why specific alerts were generated, investigate incidents thoroughly, and make informed decisions about security responses.

Regulatory frameworks such as the EU's General Data Protection Regulation (GDPR) and emerging AI governance standards require that automated decision-making systems provide meaningful explanations for their outputs. This requirement often conflicts with the use of the most effective AI models, which tend to be the least explainable.

To address this challenge, various explainable AI (XAI) techniques have been developed. Table 2 compares eight popular XAI methods used in cybersecurity, highlighting their strengths and trade-offs in terms of explanation quality, implementation complexity, and performance impact. These techniques are vital for ensuring transparency in AI-driven decisions, which is especially important in environments governed by regulatory requirements. *Note: The techniques presented in this table are derived from (Zhang, 2022).*

Table 2: Comparison of Explainable AI Techniques Used in Cybersecurity Applications

Technique	Cybersecurity Application	Explanation Quality	Performance Impact
LIME (Local Interpretable Model-Agnostic Explanations)	Malware classification Network anomaly detection	★★★★ Implementation: Medium	Low (5-10% overhead)
SHAP (SHapley Additive exPlanations)	Feature importance in threat detection Risk scoring explanations	★★★★★ Implementation: Medium-High	Medium (10-15% overhead)
Attention Mechanisms (Model Specific)	Deep learning-based intrusion detection Sequential pattern analysis	★★★★ Implementation: High	Low (2-5% overhead)
Decision Trees/Random Forest (Intrinsically Interpretable)	Rule-based threat classification Policy violation detection	★★★★★ Implementation: Low	None
Linear Models with Regularization (Intrinsically Interpretable)	User behavior scoring Risk assessment	★★★ Implementation: Low	None
Counterfactual Explanations (Model Agnostics)	"What-if" threat scenario analysis Security policy optimization	★★★★ Implementation: High	Medium (15-20% overhead)

Technique	Cybersecurity Application	Explanation Quality	Performance Impact
Gradient-based Attribution (Model Agnostics)	CNN-based malware detection Image-based threat analysis	★★★ Implementation: Medium	Low (3-8% overhead)
Rule Extraction (Post-Hoc)	Converting neural networks to rules Compliance reporting	★★★★ Implementation: High	High (20-30% overhead)

Legend:

- **Explanation Quality**: ★ = Poor, ★★★★★ = Excellent
- **Implementation Complexity**: Low = Minimal coding/setup, Medium = Moderate configuration or customization required, High = Extensive development required
- **Performance Impact**: Computational overhead when generating explanations

Key Considerations for Selection:

- **For Real-time Applications:** Choose LIME, Attention Mechanisms, or intrinsically interpretable models
- **For Regulatory Compliance**: Prioritize SHAP, Decision Trees, or Rule Extraction
- **For Security Team Training:** Focus on Decision Trees, SHAP, or Counterfactual Explanations
- **For Automated Reporting**: Implement SHAP or Rule Extraction with standardized output formats

Resource Requirements and Implementation Challenges

Implementing enterprise-grade AI security systems requires substantial computational resources, specialized expertise, and ongoing operational costs. High-performance GPUs or specialized AI accelerators are necessary for real-time processing of large-scale security data.

Cost analysis from Gartner indicates that organizations implementing comprehensive AI-powered security platforms invest an average of $2.4 million in initial implementation costs, with ongoing operational expenses averaging $600,000 annually for large enterprises, as part of broader security spending projected at $213 billion in 2025 (Gartner Security Research, 2025).

The environmental impact of AI in cybersecurity is a growing concern. Data centers powering AI systems consumed 4.4% of U.S. electricity in 2023, a figure that could triple by 2028, contributing to significant carbon emissions and water usage for cooling (IEEE, 2025).

Critical Skills Gap

The intersection of AI and cybersecurity requires expertise in both domains, creating a significant shortage of qualified professionals. The (ISC)² Cybersecurity Workforce Study 2025 reports a global cybersecurity workforce gap of 4.8 million professionals (a 19% increase from 2024), with AI-cybersecurity specialists representing the fastest-growing and most difficult-to-fill positions (ISC², 2025).

Universities and training organizations are struggling to develop curricula that adequately prepare professionals for AI-cybersecurity roles, which require understanding of machine learning algorithms and their security implications, adversarial ML attack and defense techniques, AI system deployment and operational security, ethical AI development and bias mitigation, and regulatory compliance for AI systems.

5. Emerging Trends and Future Considerations

The future of AI in cybersecurity will be characterized by several key trends that will reshape how organizations approach digital defense and how threat actors conduct attacks.

Zero-Trust Architecture Enhanced by AI

Zero-trust security models, which assume no inherent trust for any user or device, are being significantly enhanced by AI capabilities. Modern zero-trust implementations utilize machine learning to provide dynamic risk assessment, continuous authentication, and adaptive access controls.

AI-enhanced zero-trust systems analyze real-time behavioral patterns, device characteristics, network conditions, and threat intelligence to calculate dynamic trust scores for every access request. These scores determine appropriate security controls, from seamless access for low-risk scenarios to multi-factor authentication and additional verification for high-risk situations.

Microsoft's implementation of AI-powered zero-trust architecture has demonstrated 99.9% reduction in successful credential-based attacks, 80% decrease in user friction for legitimate access requests, 60% improvement in insider threat detection accuracy, and $50 million annual cost savings through automated access management (Microsoft Digital Defense Report, 2024).

Quantum Computing Implications

The development of practical quantum computers poses both opportunities and challenges for AI-powered cybersecurity. Quantum computing threatens current cryptographic standards, potentially rendering existing encryption methods obsolete within the next 10-15 years.

However, quantum machine learning algorithms may provide exponential improvements in certain AI cybersecurity applications through quantum neural networks for pattern recognition in network traffic, quantum optimization algorithms for resource allocation in security systems, quantum-enhanced threat intelligence analysis and prediction, and post-quantum cryptographic algorithm development and implementation.

The National Institute of Standards and Technology (NIST) finalized post-quantum cryptography standards in August 2024, including algorithms like CRYSTALS-Kyber for quantum-resistant encryption. Organizations were advised to begin migration planning immediately (NIST, 2025).

Federated Learning for Collaborative Threat Intelligence

Federated learning enables multiple organizations to collaboratively train AI models without sharing sensitive data directly. This approach addresses privacy and competitive concerns while enabling the development of more comprehensive threat detection capabilities.

Implementation offers several advantages including preservation of data privacy and regulatory compliance, access to larger and more diverse training datasets, reduced individual organizational training costs, and improved model robustness through diverse data sources.

The Financial Services Information Sharing and Analysis Center (FS-ISAC) has initiated a federated learning pilot program involving 25 major banks, demonstrating 40% improvement in fraud detection accuracy compared to individual institution models (FS-ISAC, 2025).

AI Governance and Regulatory Evolution

The regulatory landscape for AI in cybersecurity is rapidly evolving, with new frameworks emerging globally. The EU AI Act, effective 2024, classifies AI systems used for cybersecurity as "high-risk" applications requiring extensive documentation, testing, and compliance measures.

Key regulatory trends include mandatory algorithmic impact assessments for AI security systems, requirements for explainable AI in critical security decisions, auditing standards for AI model training data and processes, and incident reporting requirements for AI system failures or compromises.

6. Best Practices and Strategic Recommendations

Organizations should adopt a structured, phased approach to implementing AI in cybersecurity programs, following established frameworks while addressing unique organizational requirements and risk profiles.

Comprehensive Implementation Framework

This framework assumes that the organization has a basic level of cybersecurity infrastructure in place, including threat detection systems, security information and event management (SIEM), and an awareness of AI concepts. The phases described below are based on the assumption that AI tools will be introduced gradually, starting with controlled environments before scaling up across critical functions. The timeframes mentioned are averages and may vary depending on organizational size, AI readiness, and available resources.

Phase 1: Foundation and Assessment (3-6 months)

- Conduct comprehensive risk assessment of current security posture
- Evaluate data quality, availability, and governance capabilities
- Assess organizational AI readiness and skill requirements
- Establish AI governance structures and decision-making processes

Phase 2: Pilot Implementation (6-12 months)

- Deploy AI systems in limited, controlled environments
- Focus on high-impact, low-risk use cases such as threat intelligence analysis
- Establish baseline performance metrics and success criteria
- Develop staff training programs and change management procedures

Phase 3: Scaled Deployment (12-24 months)

- Expand AI implementation to critical security functions
- Integrate AI systems with existing security infrastructure
- Implement comprehensive monitoring and performance management
- Establish continuous improvement and model updating procedures

Human-AI Collaboration Optimization

Effective AI implementation requires thoughtful integration of human expertise and machine capabilities. Research from Stanford's Human-Centered AI Institute demonstrates that optimal cybersecurity outcomes occur when AI systems augment rather than replace human analysts.

AI excels at pattern recognition, data processing, routine analysis, and rapid response, while humans excel at strategic thinking, contextual analysis, ethical judgment, and creative problem-solving. Successful collaboration models include human-in-the-loop approaches where AI provides recommendations and humans make final decisions, human-on-the-loop systems where AI operates autonomously with human oversight and intervention capability, and human-over-the-loop implementations where AI systems operate independently with periodic human review.

To enhance usability, AI security tools should prioritize human-centric design, including intuitive interfaces and clear visualizations. A 2025 SANS Institute survey found that 45% of analysts cite complex AI dashboards as a major barrier to effective use, underscoring the need for user-friendly designs (SANS Institute, 2025).

Ethical AI Deployment and Governance

Organizations must establish robust ethical frameworks for AI deployment in cybersecurity, addressing privacy and civil liberties through data minimization principles, consent and notification requirements, protection of employee and customer privacy rights, and transparency in AI decision-making processes.

Fairness and non-discrimination require regular bias testing and mitigation procedures, diverse and representative training datasets, equal treatment across demographic groups, and appeal processes for AI-driven security decisions. Accountability and responsibility demand clear ownership for AI system decisions, comprehensive audit trails and decision logging, regular performance reviews and ethical assessments, and incident response procedures for ethical violations.

Mitigating AI Risks in Cybersecurity: Addressing Key Concerns

As organizations integrate AI into their cybersecurity infrastructure, it is critical not only to focus on the advantages but also to address the inherent risks associated with AI systems. Many of the concerns raised earlier, such as adversarial attacks, bias in machine learning models, and explainability challenges, must be actively managed to ensure the effectiveness and trustworthiness of AI-powered security solutions.

Here are key strategies organizations should employ to mitigate these risks:

- **Adversarial Attacks**: Machine learning models can be manipulated to lead to incorrect predictions or compromised defenses. To combat this, organizations should incorporate **adversarial training**, a process where AI models are intentionally exposed to adversarial inputs during training to improve their robustness. Additionally, implementing defensive techniques, such as **gradient masking** and **data augmentation**, can help strengthen AI models against potential threats. Regular penetration testing and vulnerability assessments should also be conducted to identify and patch any exploitable weaknesses.
- **Bias and Fairness**: Algorithmic bias remains a significant challenge, particularly when using AI models to analyze large, diverse datasets. Organizations should adopt strategies such as **bias detection frameworks** and use diverse, representative datasets during model training. Regular **bias audits** should be performed to ensure AI systems do not unfairly target certain groups or demographic categories. The integration of explainable AI (XAI) methods, such as SHAP and LIME, can also help uncover potential biases by providing transparency into how models arrive at decisions.
- **Explainability and Transparency**: In the cybersecurity context, AI models must provide explanations for their decisions, especially when they detect potential threats. Organizations should prioritize explainable AI (XAI) techniques, ensuring that security analysts can understand the reasoning behind AI-driven alerts and actions. This can be achieved by utilizing tools like decision trees, LIME, and

SHAP, which offer insights into how a model processes inputs and produces outputs. By enhancing transparency, organizations can build trust in their AI systems and improve the overall decision-making process.

- **Continuous Monitoring and Model Refinement**: The threat landscape is dynamic, and AI models must be continually updated to remain effective. Organizations should establish processes for ongoing model evaluation and retraining, ensuring that their AI systems adapt to new attack strategies and emerging threats. Additionally, implementing a continuous feedback loop where human analysts can provide input on AI alerts and refine models over time is crucial to improving system performance.
- **Ethical Considerations and Regulatory Compliance**: AI deployment in cybersecurity must adhere to ethical guidelines and comply with relevant regulations, such as the GDPR or CCPA. Establishing AI governance frameworks that emphasize data privacy, accountability, and transparency is essential. Organizations should also prioritize data minimization principles and ensure that AI models operate fairly and without discrimination.

By incorporating these proactive measures, organizations can mitigate the risks associated with AI in cybersecurity and maximize the benefits of AI-powered security systems. Addressing these concerns in tandem with the implementation strategies outlined earlier will create a more robust and resilient cybersecurity infrastructure.

Conclusion

The integration of artificial intelligence into cybersecurity represents both an unprecedented opportunity and a significant challenge for organizations worldwide. According to recent reports, **approximately** 350,000-500,000 new malware threats are detected daily (Check Point, 2021; Kaspersky, 2021). As sophisticated attackers increasingly leverage AI to enhance their capabilities, traditional security approaches, such as signature-based detection, have proven inadequate for tackling the modern threat landscape (Symantec, 2020).

Organizations implementing AI-powered security systems demonstrate measurable improvements in threat detection speed, accuracy, and cost-effectiveness. However, success requires careful attention to the limitations and risks inherent in AI systems, including algorithmic bias, adversarial attacks, explainability challenges, and substantial resource requirements.

The future will be defined by an accelerating arms race between AI-powered attacks and AI-enhanced defenses. Organizations that invest in understanding and implementing AI security technologies today, while carefully managing their risks and limitations, will be better positioned to defend against sophisticated threats.

Critical success factors include strategic implementation following structured frameworks that balance innovation with risk management, human-AI integration developing collaboration models that optimize combined capabilities, continuous evolution establishing processes for ongoing adaptation to evolving threats and technological capabilities, and ethical governance implementing frameworks that ensure AI systems operate fairly, transparently, and in compliance with regulatory requirements.

The organizations that thrive will be those that view AI not as a silver bullet for cybersecurity challenges, but as a powerful tool requiring careful implementation, continuous refinement, and strategic integration with human expertise. The stakes are high, but so are the opportunities for those who can successfully navigate this complex and rapidly evolving landscape.

As the landscape of AI and cybersecurity continues to evolve, staying informed is crucial for organizations aiming to implement and adapt AI-driven security measures. For up-to-date guidance, case studies, and expert insights, consider subscribing to *Lights On*, where I share the latest research, trends, and best practices.

About the Author

Farida Khalaf *is a data engineer, certified in threat intelligence and cybersecurity analytics, and a global affairs expert. Her interdisciplinary* *career combines a BA in Economics with two master's degrees in International Finance and Global Affairs.*

Farida has extensive experience collaborating with the UN and leading large-scale international grant projects. Her work provides a unique and pragmatic perspective on the future of AI in agriculture, bridging technical data expertise with real-world global policy. She is based in NYC, USA.

Chapter 20

From Privacy and Cybersecurity to AI Governance

By Soribel Feliz

Intended Audience: *People with experience in privacy or cybersecurity who are interested in moving into AI governance.*

IF YOU WORK in privacy, cybersecurity, or Governance, Risk and Compliance (GRC), you've probably read an artificial intelligence (AI) governance job description and thought: *This makes no sense.* Companies are asking for data scientists with policy backgrounds, engineers who understand ethics, and compliance professionals who can also code. In other words: unicorns.

The problem is not that AI governance is impossible. The problem is that organizations don't yet know how to describe the work. What they are actually looking for is governance. And governance is not new.

Most AI governance work is not about algorithms. It's not even about understanding neural network architecture diagrams. It's about designing the right decision-making mechanisms, clarifying accountability, understanding data processing purpose, and making sure people are not harmed. These terms should sound familiar. This is what privacy professionals do every day.

AI Governance Is Not a Brand-New Discipline

AI governance is not a brand-new discipline pulled out of thin air. It is an extension of compliance work that already exists. The difference is the system being governed.

AI systems process data, make decisions, and create risk. That puts them squarely in the territory of people who understand accountability, documentation, controls, and oversight. The professionals best positioned to do this work often underestimate their own relevance because the language has changed, not the substance.

AI governance and privacy are symbiotic. Not adjacent or related, but absolutely symbiotic. Most of what organizations call "AI governance" is really decision-making, accountability, purpose discipline, and harm prevention. A lot of that is privacy work. One cannot govern AI without understanding how data is collected, labeled, retained, shared, and repurposed across the lifecycle.

Global regulators and data protection authorities constantly reiterate that existing privacy and data protection rules apply to AI systems. The EU AI Act explicitly layers on top of General Data Protection Regulation (GDPR). The Federal Trade Commission (FTC) has made clear that existing consumer protection laws cover algorithmic harms. This makes the integration of privacy and AI governance even less optional.

The Compliance Mindset Maps Directly to AI Governance

If you have spent your career in compliance, you already know how to operate in ambiguity. You know how to take a loosely defined regulatory expectation and turn it into a process that can be repeated, monitored, and defended. AI governance requires the same instincts. You ask questions like:

- Who owns this system?
- What data is it using?
- How do we know it is working as intended?
- What happens when it fails?
- How do we explain our decisions to regulators or leadership?

These are operational questions, not abstract ethical questions. They are the same questions you asked as a privacy professional when GDPR came into force, when new cybersecurity standards were introduced, or when your organization rolled out a new system that touched sensitive data. The tools may be different. The labels may be new. The structure of the work is familiar.

Think of a typical privacy workflow. You receive a new product intake request. You are not a product owner, so you ask questions:

- What data is collected?
- How does it flow? Who accesses it?
- How long is it stored?
- How are individuals informed?
- What happens if someone exercises their rights?
- How are risks mitigated?

In the world of AI, you ask similar questions, but about the model itself:

- What is this model designed to do?
- What training data was used, and where did it come from?
- Who has access to the model outputs?
- How are decisions explained to affected individuals?
- What happens if the model produces discriminatory outcomes?
- How do we handle data subject rights when decisions are automated?

The framework is identical. The subject matter has shifted from databases and applications to models and algorithms, but the governance discipline remains constant.

Risk Assessment Is Not New

One of the most misunderstood aspects of AI governance is risk assessment. People often assume this requires deep technical expertise. In reality, it requires the ability to identify potential harm, assess likelihood and impact, and put controls in place.

Privacy professionals have been doing this for years. An AI risk register looks very similar to any other risk register. You identify risks such as bias, data misuse, model drift, or lack of transparency. You document mitigation measures. You assign ownership. You define monitoring mechanisms. You revisit the assessment as the system evolves. This work is practical, repeatable, and auditable. Those are compliance strengths.

Consider a practical example: Your organization wants to deploy an AI-powered resume screening tool. A privacy professional approaches this the same way they would approach any high-risk processing activity:

Identify the purpose: What problem is this solving? Why is automation necessary?

Map the data: What data feeds the model? Is it representative? Does it contain protected characteristics or sensitive information?

Assess the impact: Who could be harmed if this system fails? What are the consequences of a false positive or false negative?

Design controls: How will we test for bias? What human oversight exists? How do we document decisions?

Plan for accountability: If someone challenges a decision, can we explain it? Can we correct it?

This is a data protection impact assessment applied to an AI system. You already know how to do this.

Documentation and Accountability Still Matter

AI systems do not remove the need for documentation. They increase it. Organizations deploying AI need to be able to show how decisions were made, who approved them, what data was used, and how outcomes are monitored. They need records that can withstand internal review, external audits, and regulatory scrutiny.

If you have built audit trails, maintained compliance documentation, or prepared an organization for regulatory exams, you already understand what good documentation looks like. AI governance depends on that same discipline. Regulators are not asking for a detailed explanation of how a model works at the mathematical level. They are asking who is responsible, how risk is managed, and how harm is prevented. That is accountability, not code.

The narrative around AI has been dominated by engineering voices and research. When entering through this door, it is easy for a privacy professional to feel like a guest in the AI space. But the engineers are often just as lost when it comes to governance questions. They can build the model, but they don't know how to operationalize data protection requirements, navigate regulatory obligations, or communicate with non-technical stakeholders about risk. That's where you come in.

What Privacy Professionals Bring: The Discipline of Inquiry

Privacy professionals operate with a fundamental skepticism about how systems handle data and impact people. When someone in the organization says, "We need to collect this information," the immediate response is: "For what purpose? Is it necessary? What's the minimum we actually need?"

This instinct to question assumptions rather than accept them is precisely what AI governance requires. When a team proposes using a particular training dataset because "it's industry standard" or "everyone uses it," someone needs to ask about its provenance, representativeness, and fitness for purpose. When engineers report that a model achieves 95% accuracy, someone needs to ask: "Accurate for whom? What happens to the 5% who are misclassified? What are the consequences of being wrong?"

These questions don't really require deep technical knowledge, despite what recent AI compliance job descriptions tell you. They require the governance discipline that privacy professionals have been practicing for years: refusing to take *"because we've always done it this way"* as sufficient justification for processing decisions that affect people.

The most effective AI governance comes from bringing a different lens to the conversation: one focused on purpose, necessity, proportionality, and impact. Privacy professionals already think this way about every new system, every data flow, every processing activity. AI systems are no exception.

What Technical Literacy Actually Means

In this current AI adoption age, privacy professionals and AI governance leads both need to become bilingual in two languages: legal rules and technical reality. That means being able to identify proper versus improper data use, map privacy obligations onto model development and deployment, and operationalize controls that engineers can actually implement.

But this doesn't require becoming an engineer. Privacy professionals already navigate this dynamic with other technical domains. You don't perform penetration testing yourself, but you understand what a data breach is, how encryption works conceptually, and what questions to ask about security controls. You don't write database queries, but you understand data flows well enough to assess retention and deletion risks. The same principle applies to AI systems.

For AI governance work, you do not need to:

- Write code or build machine learning models from scratch
- Understand the mathematical foundations of algorithmic frameworks
- Fine-tune hyperparameters or optimize neural networks
- Debug algorithms or conduct technical audits independently

You do need to:

- Understand how AI systems work at a conceptual level (inputs, processing, outputs)
- Recognize common failure modes (bias, drift, overfitting, data leakage)
- Ask informed questions about model design, training data, and deployment contexts
- Translate between technical teams and business stakeholders
- Know when to escalate technical concerns and who to escalate them to

Some technical work in AI governance requires specialized expertise: validating fairness metrics, implementing privacy-preserving techniques, checking fine-tuning guardrails. The question is not whether you can do all of this yourself. The question is whether you can coordinate the people who can, ensure accountability is clear, and confirm that governance requirements are being met. That's the same coordination work privacy professionals already do across security, legal, engineering, and product teams.

The Reality of Cross-Functional Coordination

Privacy work has never been solitary. It requires constant coordination with security teams, engineering, legal, product management, and external vendors. Privacy professionals routinely ask the right questions and ensure someone accountable provides the answers. They don't need to conduct penetration testing themselves, but they need to understand what security controls exist and who owns them. They don't need to write code, but they need to understand data flows, both across teams and within organizations, well enough to identify risks.

AI governance operates identically. Someone in the organization needs to understand model drift and validate fairness metrics. Someone needs to implement privacy-preserving techniques and test for algorithmic vulnerabilities. That someone doesn't have to be you, but you need to know who it is, what they're accountable for, and how their work integrates with governance requirements.

This coordination model is already second nature to privacy professionals. How many times have you led projects specifically to fill gaps that no single team owned? Privacy shortcomings in an HR system get addressed by convening engineering and people ops. Data transfer security for a migration project requires conversations with security and infrastructure teams. The same collaborative framework applies to AI systems.

Privacy professionals also serve as translators between technical and non-technical stakeholders, a role that becomes even more critical in AI governance. Explaining GDPR requirements to a product manager who wants to ship a feature, or translating *"because Legal said so"* into operational controls that engineers can actually implement, requires the ability to move between different vocabularies and perspectives without losing fidelity to either side. AI governance demands this same translation capability: helping engineers understand why regulators care about explainability, or helping legal teams understand why certain technical constraints make perfect transparency impossible.

When positioned as governance partners rather than compliance gatekeepers, privacy professionals typically find they're actively sought out by technical teams who recognize that rigorous oversight prevents costly rework, reduces legal exposure, and builds user trust. The best technical teams understand that governance makes their work sustainable, not slower.

Incident Response Does Not Disappear

When an AI system behaves unexpectedly, the response should not be improvised. It should follow a defined process. Compliance and security professionals already know this playbook: Detect the issue. Assess impact. Contain the risk. Fix the problem. Document what happened. Adjust controls to prevent recurrence. Whether the incident is a data breach or an AI system producing harmful outcomes, the response framework remains the same. AI governance does not replace incident response. It extends it.

Consider what happens when a content moderation algorithm starts incorrectly flagging legitimate speech, or when a credit scoring model begins producing disparate outcomes for protected groups. The technical team can follow a similar playbook: diagnose the model malfunction, but someone needs to assess legal exposure, communicate with affected individuals, coordinate with regulators if necessary, and implement systemic fixes. That's governance work.

What Regulators Actually Expect

There is a persistent assumption that AI governance requires organizations to explain models in ways that are unrealistic or overly technical. In practice, regulators care about governance structures, accountability, and evidence of control. They want to know:

- Who approved this system?
- What safeguards are in place?

- How do you monitor ongoing performance?
- What happens when something goes wrong?

These are governance questions. They are questions compliance professionals are trained to answer.

Privacy and AI governance should be deeply integrated into every AI team, not to slow innovation, but to reduce rework, prevent foreseeable harm, and increase customer trust. Trust is not a soft metric. It translates to product adoption, retention, and return on investment. Organizations that treat AI governance as a compliance checkbox rather than a strategic function consistently underperform on all these measures.

Repositioning Your Skill Set

The biggest barrier for compliance professionals is not lack of capability. It is positioning.

Many people assume they need to become engineers or data scientists to work in AI governance. That is not true. What is required is enough technical literacy to ask informed questions and understand tradeoffs.

Your value is not in building models. Your value is in governing them. That means translating your experience into language that AI teams and hiring managers understand. Risk management becomes AI risk management. Control frameworks become AI governance frameworks. Monitoring becomes model oversight. Stakeholder engagement becomes cross-functional collaboration on responsible AI. The work is not new. The context is.

When you apply for AI governance roles or pitch yourself for internal transitions, don't lead with what you don't know. Lead with what you do: *"I've spent five years implementing privacy compliance programs that required coordinating across engineering, legal, and product teams. I've conducted hundreds of risk assessments on high-stakes systems and translated regulatory requirements into operational controls. I'm now applying that same framework to AI systems."* That's not a career pivot. That's a natural evolution.

A Practical Path Forward

If you are considering this transition, start where you are. Treat AI tools in your organization like any other high-risk system. Ask who owns them. Ask what data they use. Ask how risk is assessed and monitored. Offer to help formalize processes that already exist informally.

Volunteer to join AI governance working groups or risk committees. Many organizations are standing these up without clear leadership. This is your opportunity to demonstrate value before the role is even defined.

Learn enough about AI to understand its failure modes, not to become an engineer. Take an introductory course on machine learning concepts. Read case studies about algorithmic failures. Follow practitioners who are doing this work (not just academics theorizing about it). Consider pursuing certifications like the IAPP's Artificial Intelligence Governance Professional (AIGP) credential, which provides structured learning in AI governance principles (IAPP, 2026). Or you may take a foundational AI Governance course like the one I produced for LinkedIn Learning. (For links to these courses and other chapter resources, see the book website.)

Build demonstrable skills rather than just collecting credentials. Create a portfolio that shows your ability to conduct AI risk assessments, design governance frameworks, or facilitate cross-functional collaboration around ethical issues. One successful career changer I know created a detailed case study analyzing bias in a public dataset and proposing mitigation strategies, then used this case study to demonstrate her capabilities during interviews.

Most importantly, recognize that AI governance is operational work. It lives at the intersection of risk, accountability, and execution. That is where compliance professionals already operate.

The Invitation

My message to every privacy colleague who has quietly wondered if they belong in conversations about AI is this: **Yes, you do.**

You are not a guest or an outsider. You are not unqualified. You can protect humans from data misuse, algorithmic risk, and model drift. If you can explain complex privacy risks to business leaders, you can explain AI risk too. If you have survived a round of privacy risk assessment comments from legal, product, and engineering, trust me, you already have the patience and resilience needed for AI governance.

AI is exciting. It is collaborative. It is complex. It is the next chapter. And we privacy and cyber professionals are natural leaders in this space, if we stop waiting for permission and start claiming the expertise we already have.

Closing Thought

AI governance does not belong only to technologists or theorists. It belongs to people who know how to build systems that hold up under scrutiny. If your career has trained you to ask hard questions, document decisions, manage risk, and protect people from harm, you are already doing the work. The label has changed. The foundation has not.

The field needs people who understand that governance is about building sustainable systems that can scale without causing preventable harm, not about saying "no" to innovation. It needs people who can sit in a room with engineers, lawyers, product managers, and executives and ensure everyone is speaking the same language about risk and responsibility. That's you. You just need to recognize it.

About the Author

__Soribel Feliz__/USA is the founder of __Algorithms Are Personal__, an AI-governance company. Soribel was an AI governance lead at the Dept. of Homeland Security and former Senior Tech Policy Advisor to the U.S. Senate. With over 12 years of experience spanning federal government, Congress, and technology companies, including Microsoft and Meta, she specializes in translating complex AI regulations into practical compliance frameworks. A former diplomat, Soribel helps organizations navigate the intersection of technology policy and regulatory risk management.

Chapter 21

Empathy by Design

How AI and UX Shape a More Inclusive Digital World

By Blessing Okpala, PhD

Intended Audience: *People who work in the field of user experience and product development, such as UX designers, AI researchers, and product leaders.*

Introduction: Why Empathy Still Matters in a Machine-Led World

GROWING UP, I was my mother's handbag. Wherever she went, such as market runs, errands, and visits to the neighbours, I went too, carrying bags, holding change, and learning something far deeper than chores. I didn't know it then, but I was getting a masterclass in empathy. Empathy refers to the ability to understand another person's feelings, experiences, and perspective so that their needs and context inform how systems and interactions are designed (Rahaman et al.,2023).

Market days were loud and crowded, bursting with colors, smells, and sounds. What captured my attention most weren't the yams or palm oil, but the people navigating the space: visually impaired individuals guided by loved ones, wheelchair users being pushed by friends or siblings, and deaf people holding handwritten "Please help me" signs. These people weren't charity cases; they were living, working, and participating fully in society but relying on others to bridge gaps in the environment that wasn't built for them.

I recall a boy guiding his visually-impaired father through the stalls, narrating each step and translating obstacles into actionable cues. The father experienced the market, not directly but through his son's words. Similarly, a woman in a wheelchair made her own choices while her sister became her guide through uneven paths. In those moments, I realized that when spaces aren't designed for everyone, **people become the interface**, filling gaps that should have been addressed by thoughtful design.

Figure 1: Market scene showing accessibility challenges. Source: Image generated with Google Gemini, prompted by the author, January 2026.

Years later, as I transitioned into UX design, those lessons came rushing back. Digital products often fail people in small but meaningful ways; for example, voice assistants mishear accents, navigation apps ignore accessibility needs, and AI chatbots unable to interpret limited literacy. Like the market without guides, these systems operate logically but lack human understanding. Technology is only as compassionate as the design behind it (ACM, n.d.). Throughout this chapter, inclusive AI is framed as an ethical commitment to dignity and autonomy, ensuring people are respected, empowered, and able to participate fully in digital experiences.

This chapter explores how UX design and AI can restore that human touch at scale. UX brings insight into real human behavior, such as what frustrates, delights, or confuses users, while AI enables scalability, personalization, and adaptability to diverse needs. But without empathy, AI risks being exclusionary, impersonal, and even harmful.

We will examine a dual responsibility shared by UX designers, AI researchers, product leaders, and the organizations that design, build, and deploy intelligent systems.

First, **AI can create inclusive outcomes**, for example, tools like Seeing AI (Microsoft) help visually impaired users "see" the world, while Project Euphonia (Google Research) helps AI understand atypical speech.

Second, **AI systems themselves must be inclusive**, ensuring that people of all abilities, languages, cultural contexts, and literacy levels can interact meaningfully. A beautifully-coded AI is meaningless if users cannot navigate, speak to, or understand it. Academic research reinforces this principle, while highlighting that exclusion often arises when designers assume users share their abilities and biases (Holmes, 2018). Human-centered AI design addresses this mismatch, making technology accessible, usable, and empowering for all.

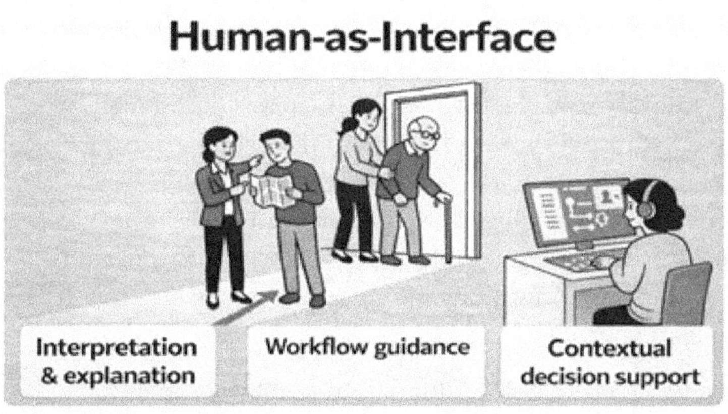

Figure 2: Conceptual Diagram of the Human-as-Interface Concept. Source: Image generated with OpenAI's ChatGPT (DALL.E), prompted by the author, January 2026.

By the end of this chapter, practitioners, students, and general readers engaged with AI and digital design will see how the fusion of UX and AI can transform technology into something that feels human again: systems that don't just work but understand. Just as people guide their loved ones through markets, empathy in design ensures everyone has a voice, a choice, and a seat at the digital table. AI can play that same role if we design it to do so.

The Intersection of UX Design and Artificial Intelligence

When I first encountered AI-powered design tools, I felt a familiar tension: the promise of technology colliding with the messiness of human need. But watching a blind student use Microsoft's Seeing AI to read a textbook independently, I understood something profound: User Experience (UX) design and Artificial Intelligence (AI) aren't competing forces and are no longer separate worlds. They have become intertwined disciplines, shaping how people experience technology, with each amplifying the other's strengths. The relationship between UX design and artificial intelligence represents a symbiotic evolution rather than a technological takeover.

While UX design focuses on human behavior, empathy, and context, AI contributes the ability to adapt, predict, and personalize at scale (Bevilacqua et al., 2025; Rahaman et al., 2023). When the two converge thoughtfully, they create systems that feel not only intelligent but also deeply human (Shrivastava, 2025). As system developers rapidly embed AI into everyday platforms and sectors like healthcare and education, its power to analyze complex data for personalized solutions must be balanced by thoughtful UX design, ensuring these systems are transparent, fair, and accessible to everyone, thereby making efficient technology more understandable and trustworthy (Maslej et al., 2025; Ho, 2025).

Consider Netflix's recommendation engine, where AI analyzes viewing patterns, while UX design ensures those suggestions feel personal rather than invasive (Cherukuri, 2024). As highlighted by researchers, 80% of content watched on Netflix comes from AI recommendations (Liu et al., 2024), yet users feel they are discovering content themselves. This bidirectional enhancement, whereby UX is making AI trustworthy, while AI is making UX adaptive, defines the new paradigm.

In education, adaptive learning platforms demonstrate this synergy beautifully. The next generation of adaptive learning platforms should leverage Generative AI and inclusive design to dynamically customize all aspects of learning in real-time, creating equitable and effective environments for students with disabilities (Strielkowski et al., 2025). The AI handles individualization at scale; the UX design ensures each interaction feels supportive rather than surveillance-based. Yet this partnership demands careful choreography. AI is transforming Human-Computer Interaction (HCI) by enabling personalized, adaptive user experiences, thus necessitating a strong focus on ethical design and bias mitigation for responsible implementation (Kerdvibulvech & Jiang, 2025).

While Human-Centered Artificial Intelligence (HCAI) is focused on designing AI systems that prioritize human needs, satisfaction, and trustworthiness, especially in fields like healthcare, the field currently lacks standardized methodologies and suffers from vague guidelines that make practical implementation challenging (Bevilacqua et al., 2025). The core requirement for successful AI integration is continuous mutual adjustment: UX designers must understand AI's technical constraints, and AI developers must appreciate real-world human interaction patterns to achieve bidirectional alignment (Lee & Kahng, 2025; Shen et al., 2025).

The most exciting development would be interfaces that respond instantly to user behaviour and environmental context while adapting carefully to emotional cues, since those signals aren't the same for everyone (Hadimlioglu & Linga, 2025). These real-time adaptive interfaces represent human-centered design at its most sophisticated state, using machine learning to create experiences that feel simultaneously powerful and personal.

As one framework notes, the field is deepening its focus on human-centered AI by moving toward human-AI symbiosis, where technology works alongside people to augment and not replace human capabilities (Bevilacqua et al., 2025). This intersection isn't just about better products but about expanding who can participate in digital life.

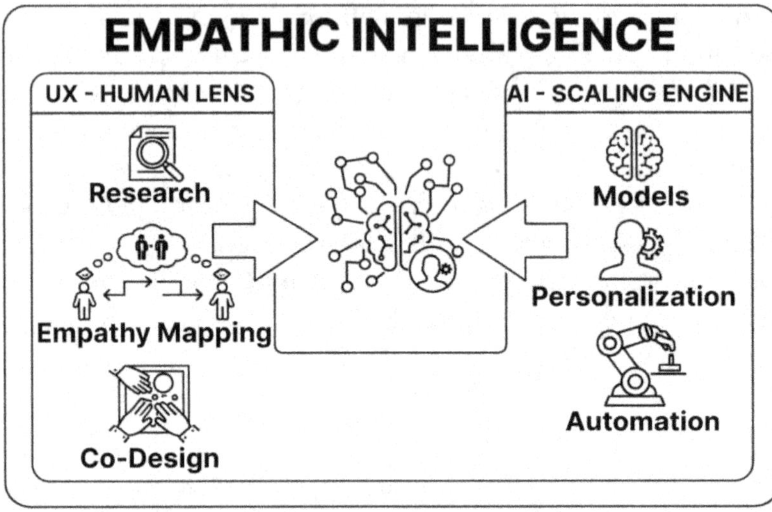

Figure 3: UX + AI → Empathic Intelligence. Source: Image generated with OpenAI's ChatGPT (DALL.E), prompted by the author, Jan. 2026.

Empathy as a Design Superpower

Empathy transforms design from solving technical problems to addressing human needs. Traditional design processes rely heavily on individual designer empathy, limiting understanding to small groups and making insight discovery coincidental and unrepresentative (Zhu & Luo, 2024). Human-centered design demands systematic empathy development that scales beyond individual intuition.

Kat Holmes articulates a foundational principle: "Designing with, not for, excluded communities is how we put the inclusive in inclusive design" (Holmes, 2018). Kat is an influential design executive and the author of Mismatch: How Inclusion Shapes Design. She pioneered the methodology of Inclusive Design by framing disability as a "mismatched interaction" that drives innovation. True inclusion requires learning from and collaborating with people who face exclusion. Microsoft's Inclusive Design Toolkit operationalizes this through "solve for one, extend to many" (Microsoft Design, n.d.), with a concept that exemplifies how designing for someone with one arm benefits 26,000 people with permanent conditions, while extending to anyone with a broken arm or holding a baby, reaching 21 million people.

The Xbox Adaptive Controller also exemplifies this philosophy (Microsoft Garage, n.d.). Developed through years of co-creation with gamers with disabilities, accessibility advocates, and organizations like AbleGamers Foundation and Cerebral Palsy Foundation, the controller's UX design process illustrates the disability community mantra: *"Nothing about us, without us"*. The result revolutionized accessible gaming while inspiring mainstream controller innovations, serving over 46 million U.S. gamers with disabilities (Kemp, 2024).

AI can help designers scale empathy by analyzing patterns in language and interaction, such as what users type or how they navigate, while recognizing that any inferred emotional signals can be imperfect and must be handled carefully. This lets designers spot important patterns across thousands of users at once, providing insights that still help them design for one person's specific needs (Zhu & Luo, 2024). AI-driven personalization can adapt layouts and content to individual needs, helping users with cognitive or learning disabilities through simplified interfaces and tailored navigation (Edwards, 2025).

Recent research on intellectual disabilities demonstrates concrete applications. By designing with users who have intellectual disabilities, new guidelines emphasizing tailored communication, accessible supports, and team collaboration proved effective, significantly boosting designer empathy, participant engagement, and data accuracy through the use of modified design tools (Corsetti et al., 2025).

Educational contexts reinforce empathy's centrality. Studies across global settings confirm that empathy-driven approaches ensure pedagogical relevance across diverse settings, moving design beyond technical efficiency to address learners' emotional and cognitive needs (Bader et al., 2025). Empathy isn't sentiment; it's the systematic practice of understanding lived experiences that produces genuinely inclusive, usable outcomes at scale. AI can amplify empathic design when treated as an assistive layer rather than a replacement for human judgment.

Accessibility in the Age of AI

AI-driven accessibility tools demonstrate both transformative promise and critical gaps. Microsoft's Seeing AI (Shaikh, 2023), which narrates the world for blind and low vision people, is now available on Android devices for the first time via the Google Play Store, featuring new generative AI capabilities like richer photo descriptions and the ability to chat with scanned documents.

Be My AI (Coates, 2024) provides conversational image descriptions, reading not just “soup can” but brand, ingredients, and expiration dates, enabling independent grocery shopping.

Google's Project Euphonia (Martin et al., 2025) exemplifies AI's potential for inclusive outcomes. By collecting speech samples from over 3,000 speakers with non-standard speech, including ALS, Parkinson's, and cerebral palsy, personalized Automatic Speech Recognition (ASR) models reduced word error rates significantly. This demonstrates that AI can outperform human transcribers for individuals with disordered speech. The initiative expanded globally to Spanish, French, Japanese, Hindi, and five Ghanaian languages, with open-source tools enabling developers worldwide to build localized solutions.

Yet critical gaps persist in both AI accessibility and AI's own accessibility. Systematic review of 43 academic articles (2018 to 2023) reveals predominant focus on visual impairments, with significant underrepresentation of speech, hearing, cognitive, and motor disabilities. Compounding this research gap, people with disabilities comprise only 0.5–3% of AI training data populations, further limiting the accuracy and effectiveness of AI systems for users with disabilities (Chemnad & Othman, 2024).

This concentration on visual impairments, while important, highlights the urgent need for more equitable distribution of research efforts across all disability types. Making AI tools themselves accessible proves equally challenging. Research shows AI bias against disability-related terms, with models classifying sentences as “toxic” based merely on disability terminology (Caulderwood, 2025).

Generative AI raises concerns around verifiability, training data ableism, and false promises. Azure AI Foundry's accessibility updates, such as grouping notifications to reduce cognitive overload and enhancing screen reader and keyboard navigation, demonstrate that intentional accessible design is essential even for the platforms used to build AI tools (Beatman, 2025). These updates remind us that inclusive AI depends not only on accessible products, but also on accessible tools that enable diverse teams to build them.

The dual imperative remains clear: AI must both create inclusive outcomes for users and ensure AI tools themselves work for people across abilities, languages, literacy levels, and cultural contexts.

Designing for Inclusivity Beyond Compliance

Inclusive design is more than simply meeting accessibility regulations or following design guidelines. It's a commitment to designing with diverse people and contexts in mind, and it's a smart business strategy that improves the digital experience for everyone. When accessibility is prioritized, a brand's commitment to user needs shines through, which not only builds trust but also opens doors to new markets. Ultimately, successful companies realize the simple truth that building products that everyone can use is essential. Current analysis suggests that inclusive web design continues to exhibit persistent shortcomings by prioritizing technical compliance over the actual user experience, thereby failing to fully actualize the core ethical principles of dignity and autonomy for diverse users (Lalitha, 2025).

Inclusivity in design must evolve beyond compliance checklists and accessibility mandates. While standards such as the Web Content Accessibility Guidelines (WCAG) (WCAG Working Group, 2025; WAI, 2024) remain essential baselines, they cannot alone guarantee belonging, equity, or dignity in digital experiences. True inclusivity involves designing systems that empower participation from all users, across ability, culture, language, age, and socioeconomic background, while recognizing that accessibility is not a technical endpoint but an ethical and human-centered process. True inclusivity centers on dignity, autonomy, and belonging. The Inclusive Design Principles (Swan, 2025), developed to bridge gaps that WCAG does not fully address, emphasize providing comparable experiences, offering choices, prioritizing content, and adding value. As accessibility experts note: *"We're not just talking about ramps and lifts, we're talking about dignity, autonomy, and belonging"* (BCO, 2025).

Apple's 2024-2025 accessibility innovations exemplify this philosophy. Eye Tracking enables iPad/iPhone control using only eyes, requiring no additional hardware through on-device machine learning. Music Haptics lets deaf and hard-of-hearing users experience music through iPhone's Taptic Engine with synchronized vibrations. There is also the Assistive Access designed with cognitive disability community feedback, which offers high-contrast buttons, large text labels, and simplified app combinations, reducing cognitive load while maintaining dignity. (Apple, 2024)

The business case for accessibility and UX is compelling. Forrester Research, cited by (Edwards, 2025; Bureau of Internet Accessibility, 2024), estimates that for every dollar invested in accessibility and user experience improvements, organizations may see up to $100 in return. While this return is not specific to AI-driven personalization, intelligent systems can help scale and sustain these gains by adapting interfaces to diverse needs, reducing friction, and improving usability across a wider range of contexts. In this way, AI becomes less a replacement for inclusive design and more a mechanism for extending its impact when guided by ethical, user-centered practice.

The inclusion of people with disabilities is both a moral and economic opportunity, and Microsoft demonstrates this by combining AI with an internal culture of inclusive design to develop flexible digital tools intended to empower people at global scale (CEE, 2023). This focus on inclusion drives significant business benefits, leading to higher revenue and profit margins for businesses. Designing for dignity moves beyond meeting standards to creating environments where everyone feels valued, respected, and empowered: not accommodated, but genuinely welcomed.

Ethical Tensions and Design Responsibilities

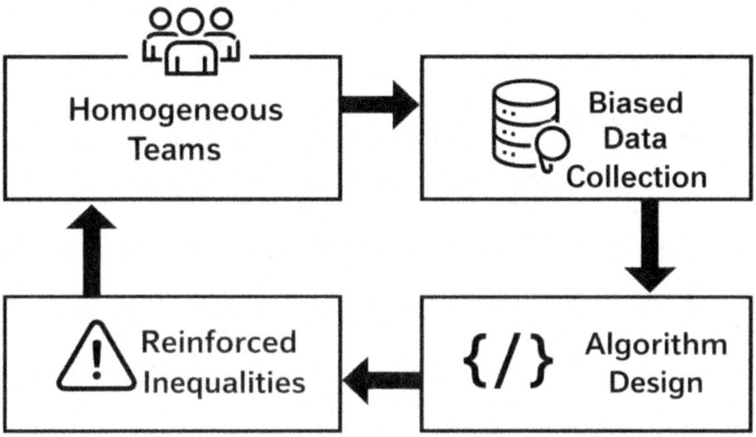

Figure 4: Bias in AI development as a feedback loop, where homogeneous teams and biased data collection reinforce algorithmic inequalities over time. Source: Image generated with OpenAI's ChatGPT (DALL.E), prompted by the author, January 2026.

The ethical landscape of AI design is complex, shaped by the biases embedded in data, algorithms, and even the teams that build them. When diversity is absent either in datasets or within design and engineering teams, AI systems risk reproducing social inequalities instead of reducing them (Xiang et al., 2025). Ethical responsibility for UX and AI practitioners goes beyond just making systems easy to use. It mandates actively addressing fairness, representation, and harm prevention by using an "ethics by design" approach to translate high-level values into concrete design requirements throughout the entire AI lifecycle (Brey & Dainow, 2024).

The consequences of bias manifest across domains; for example, studies examining contemporary healthcare AI models found that 50% demonstrated a high risk of bias, often related to absent sociodemographic data, imbalanced datasets, or weak algorithm design (Hasanzadeh et al., 2025). Examples include:

- Facial recognition systems that misclassify darker skin tones (Buolamwini & Gebru, 2018),
- recruitment algorithms that disadvantage women (Dastin, 2018), or

- language models that reproduce stereotypes (Stanford University Institute for Human-Centered Artificial Intelligence (HAI), 2025).

Each instance reveals a failure of diversity, not just in data, but in perspective.

The Stanford AI Index Report 2025 highlights this tension, warning that as AI becomes more powerful, the risks of designing systems that exclude people also grow if ethical safeguards aren't treated as a primary focus. It calls for multidisciplinary collaboration between designers, ethicists, and social scientists to ensure inclusive datasets, transparent model documentation, and participatory testing practices. UX professionals play a key role in this ecosystem by identifying usability patterns that might marginalize certain groups and ensuring ethical considerations are embedded early in design processes.

The ethics by design approach translates high-level values into concrete design requirements, embedding ethical considerations directly into the entire AI lifecycle from conception through deployment (Brey & Dainow, 2024). This requires design frameworks that help teams systematically surface potential fairness issues during early development stages, enabling proactive identification of bias before systems reach users (Zhang et al., 2023). Designers must recognize and mitigate biases in training data, ensuring fairness in decision-making processes across diverse populations through techniques that identify and address discriminatory patterns (Biondi et al., 2023).

Examples of ethical AI design include Google's Model Cards (Google, n.d.) for transparent documentation and IBM's AI Fairness 360 toolkit (Varshney, 2018), which allows teams to audit bias and improve accountability. These initiatives demonstrate how ethics can be operationalized rather than treated as abstract ideals. Ultimately, ethical UX design in AI is about responsibility with foresight: anticipating who could be excluded, challenging biased norms, and designing systems that prioritize human dignity. Inclusivity, therefore, is not a feature but an ethical commitment that defines what responsible AI truly means.

Frameworks & Strategies for Empathy-Driven AI Design

Building AI systems that are both inclusive and accessible requires intentional frameworks that embed empathy into every stage of design and development. Empathy-driven AI design is not an abstract ideal but an actionable practice grounded in participatory design, co-creation, bias auditing, and continuous testing (Bader et al., 2025). These strategies ensure that intelligent systems reflect diverse human experiences and produce equitable outcomes.

Participatory design is a foundational approach for empathy-driven innovation. It involves engaging end-users, particularly those from marginalized or underrepresented groups, throughout the design process. Participatory design turns users into co-designers (Sanders & Stappers, 2008), ensuring that AI solutions are informed by real-world perspectives rather than assumptions. Co-creation sessions, ethnographic interviews, and participatory workshops help uncover contextual needs that datasets and algorithms often overlook (Holtzblatt & Beyer, 1997; Dixon et al., 2016). This collaborative process not only reduces bias but also strengthens user trust and adoption.

Empathy mapping further operationalizes user understanding. The method visualizes users' emotions, goals, and pain points to guide feature prioritization and interaction flows (Gibbons, 2018).

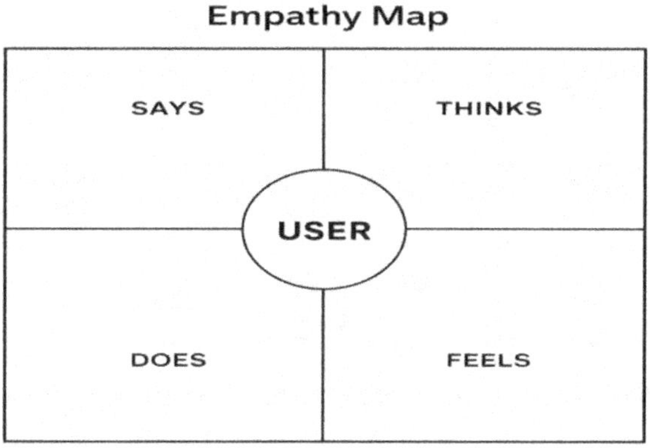

Figure 5: Empathy mapping template for capturing user perspectives across four dimensions: what users say, think, do, and feel. Source: Image generated with OpenAI's ChatGPT (DALL.E), prompted by the author, January 2026.

Empathy mapping is a critical step in developing human-centered AI frameworks, bridging the gap between cognitive insights and machine learning applications (Zhu & Luo, 2024). These frameworks align with the "empathic accountability" approach proposed by Shin (Shin, 2025), which argues that designers must continuously interpret user feedback as ethical data, ensuring systems evolve in inclusive ways.

Testing for bias is another essential strategy for inclusive AI. Tools such as IBM's AI Fairness 360 (Bellamy et al., 2019) and trained machine learning models (Mitchell et al., 2019) provide frameworks for evaluating datasets and model outputs across demographic dimensions. Integrating these audits within UX workflows allows teams to identify and correct disparities before deployment. The Fairness in Design (FID) framework, which explicitly aims to help AI software designers surface and explore complex fairness-related issues during the early design and conception phases, aligns perfectly with the concept of embedding audits before deployment (Zhang et al., 2023).

Finally, inclusive testing environments must simulate diverse user scenarios, ranging from low-bandwidth conditions to varying literacy levels and accessibility needs. Microsoft's Inclusive Design Toolkit offers practical guidance for this approach, advocating for situational testing that reveals systemic exclusion patterns (Microsoft Design, n.d.). Collectively, these strategies of participatory design, empathy mapping, co-creation, and bias testing form the backbone of empathy-driven AI. They ensure that innovation remains grounded in lived experience, leading to AI systems that are transparent, equitable, and genuinely human-centered.

The Future of Empathy at Scale

Scaling empathy requires more than larger models. It requires intentionally-designed systems that adapt to cultural norms, language variation, sensory and cognitive differences, and situational contexts (Caschera et al., 2025). Emerging research and deployments across healthcare, education, and mental health show how human-centered AI can become an empathic infrastructure if designers embed contextual intelligence, local data, and participatory processes into model lifecycles (Velush, 2025; AlMakinah et al., 2024; Thieme et al., 2023; Zia, 2024).

Example 1: In healthcare, multimodal AI copilots illustrate the potential for context-aware support. Researchers (Lu et al., 2024) have demonstrated a multimodal generative AI copilot for pathology that integrates images, text, and patient metadata to assist clinicians while preserving interpretability and workflow fit. This type of integration foreshadows tools that can surface culturally relevant information and reduce cognitive load for clinicians across settings, though evidence from clinical domains also highlights the importance of designing AI assistance that preserves clinician skill, vigilance, and autonomy.

Ambient AI systems that transcribe and summarize clinical encounters demonstrate significant productivity gains for healthcare providers as well as significant privacy risks. A prospective study by (Ma et al., 2025) confirmed that the LLM-powered scribes successfully reduced documentation time.

Example 2: In education, AI systems use generative and predictive models to customize learning. These platforms can adjust lessons, content, and feedback to fit each learner's background, existing knowledge, and unique learning style. While customizing content is powerful, recent evidence confirms that AI tools must be carefully designed to prevent harm. Design-led model tuning and local curriculum integration are necessary to avoid cultural mismatch and to preserve learner agency (Brynjolfsson et al., 2025).

This integrated approach ensures AI-powered learning is both effective and equitable. Research on human-centered AI for teamwork and training further argues that essential UX methods such as co-design, explainable feedback, and clear control affordances are prerequisites for safe, equitable, and effective adoption of AI systems in education (Chan & Hu, 2023). This intentional UX approach is necessary to ensure students' voices are heard and to prevent challenges like over-reliance and lack of understanding.

Example 3: Mental health demonstrates both the promise and the risk of empathic AI. Clinically oriented chatbots and therapeutic agents can increase access to support, particularly in under-resourced regions, but their ethical and cultural fit depends on iterative testing with diverse populations. Woebot's clinical research illustrates scalable empathic interactions, yet independent studies stress careful evaluation of language framing, cultural idioms, and escalation pathways to human care when necessary (Woebot Health, 2025; Fitzpatrick et al., 2017; Ohu et al., 2025; Brown University, 2025).

Technically, recent advances in embodied and multimodal models (e.g., PaLM-E, SAM2Act) indicate that models can fuse vision, language, and action in ways that support more natural, context-aware interactions. These capabilities, when coupled with UX research, enable interfaces that sense situational needs (Driess et al., 2023) and adapt responses accordingly (Fang et al., 2025). Governance and evaluation at scale must follow.

The Stanford AI Index 2025 stresses that organizational practices such as transparent documentation, diverse datasets, participatory evaluation, and continuous monitoring are essential for scaling empathetic AI responsibly. The report's cross-sector analysis shows that technical capability alone does not guarantee inclusive outcomes. Rather, design and governance practices determine whether empathy generalizes beyond pilot contexts.

In summary: to achieve empathy at scale, AI capabilities must be paired with human-centered design. This requires local data stewardship, participatory co-design with users, privacy-respecting multimodal sensing, and UX-driven evaluation that measures accountability. The goal isn't just more intelligent machines, but machines that are intelligible and tuned to human dignity across all cultures and contexts.

Conclusion: Designing with Dignity

The memory of those market days back home still lingers with me; for example, the visually-impaired father guided by his son's voice and the woman in her wheelchair navigating uneven ground through her sister's steady hands. Long before I had the language of UX design or artificial intelligence, those moments taught me a simple truth: when systems fail to accommodate people's needs, humans become the interface. They translate, adapt, and bridge the gaps left by poor design. Today's AI-powered world faces the same dilemma, only at a global scale.

This chapter has shown that empathy is not sentimental but structural. Research across human-AI interaction confirms that systems grounded in contextual understanding, participatory design, and inclusive datasets outperform those built on narrow assumptions [Bader et al., 2025; Stanford 2025; Sanders & Stappers, 2008]. Studies on generative AI in education and healthcare further demonstrate that when designers embed cultural nuance, linguistic variation, and accessibility needs into model lifecycles, the technology becomes more effective and equitable (Brynjolfsson et al., 2025). Similarly, the AI Index 2025 underscores that inclusive design practices like diverse evaluation cohorts, transparency, user-centered testing are as critical as algorithmic performance (Stanford 2025).

The dual responsibility explored in this chapter, where designers are responsible for how systems are built and organizations are responsible for how they are deployed, remains central. First, AI must create inclusive outcomes, from adaptive speech systems like Project Euphonia to multimodal perception tools like Seeing AI. Second, AI tools themselves must be accessible so that people of all abilities, regions, literacy levels, and cultural identities can actually use them. As recent work in accessible AI interaction emphasizes that empathy must be embedded into both the product and the process, designing with empathy is therefore not a "nice to have." It is an ethical and practical requirement for a world increasingly mediated by AI.

The future depends on systems that reduce friction instead of creating new barriers; that amplify dignity instead of diminishing agency; and that honour the full spectrum of human experience. Just as those guides in the marketplace offered clarity and support, empathic AI and UX design can ensure no one is left to navigate the digital world alone.

For ongoing writing & commentary on UX, AI, & inclusive design, readers can follow the author's newsletter at ***blessingokpala.substack.com***.

Additional professional insights and updates are shared via LinkedIn under the author's name, **Blessing Okpala**.

Acknowledgements

The author acknowledges the use of AI-assisted tools, including Google Gemini and OpenAI's ChatGPT, in the creation of original figures and visual representations in this chapter. These tools were used as assistive technologies within the author's creative and editorial process. All conceptual decisions, interpretations, and final content remain the responsibility of the author.

About the Author

Blessing Okpala, PhD, *is a Canadian digital product designer, researcher, and AI practitioner. With a background in business and doctoral research in the UK, she transitioned into technology through front-end development before specializing in UX design. Her work connects user experience, artificial intelligence, and inclusive design, with emphasis on accessibility for people with disabilities. Blessing builds AI-powered products using Lovable and prototypes with Figma MCP, GitHub Copilot, and Cursor, translating real practice into ethical, human-centered innovation.*

Chapter 22

Rapid AI Prototyping Without Technical Debt

By Elena Calvillo

Intended Audience: *People who are comfortable using an AI code generation tool to create software prototypes.*

I'VE SEEN A PATTERN that bothers me, and it's not even about AI specifically. It's about how we treat prototypes. By "prototype," I don't mean a sketch or a mock-up. I mean a functional, temporary version of a feature built specifically to validate a hypothesis before committing to full engineering.

Whether you are a product manager, an engineering leader, or a designer, you have likely seen this happen. **Teams** build something quick to test an idea. It works. Users like it. Then someone says, "Let's just ship this version," and suddenly that three-day prototype becomes the production system. A month later, **those same developers are** spending weeks untangling code that was never meant to last.

This happened to me. Lovable is an AI tool that lets you build full-stack apps just by chatting. In February 2025, I built a prototype in Lovable (using their early version) to test a conversational interface. It took maybe four hours. The code was messy. The prompts weren't versioned. But it worked well enough to show the concept. Two weeks later, I found myself trying to add a feature and realized I couldn't remember why I'd written the prompt that way. Nothing was documented. Everything was tangled together.

Sound familiar? Here's what makes this worse with AI: the tools are so good at helping us build fast that we forget we're creating technical debt at lightning speed. Unlike traditional technical debt, which is usually just messy or inefficient code that developers can eventually refactor, **AI technical debt involves "black box" logic.** It manifests as undocumented prompt behaviors, inconsistent responses, and security gaps that no one on the team understands how to fix when they inevitably break.

GitClear, a developer analytics platform that measures engineering velocity, analyzed 211 million changed lines of code in 2024 and found an 8-fold increase in duplicated code blocks. Code duplication is now ten times higher than two years ago (GitClear, 2024). This explosion in copy-pasted code suggests that while AI helps us write faster, it's actively degrading the maintainability of the codebase.

The problem isn't the AI tools. The problem is thinking that *"it's just a prototype"* means we can skip the basics that keep code maintainable.

What This Chapter Covers

This chapter gives an overview of how to test AI ideas quickly using a four-step system that prevents future headaches.

I will cover four simple habits—**naming, isolation, versioning, and decision-making**—that take minutes to set up, but save you weeks of cleanup. Then, I will share a real case study where I build a conversational prototype from scratch to demonstrate exactly how these habits work in practice. Finally, I will provide a decision framework so you know when to keep building, when to start over, and when to kill an idea before it becomes expensive technical debt.

I'm still learning this myself. I've made most of these mistakes. But at least now I know what to avoid.

Why Technical Debt Actually Matters

Ipek Ozkaya, principal researcher at the Software Engineering Institute and co-author of *Managing Technical Debt*, defines technical debt as a trade-off developers make to optimize for speed in the short term (Kruchten, 2019). It's a loan you take out against the future health of your software.

Even before the AI boom, technical debt cost the US economy $2.41 trillion annually (Consortium for Information and Software Quality, 2022). **This figure proves that technical debt is not just a minor**

engineering annoyance, but a massive financial drain that consumes budgets meant for innovation. That was the cost of fixing human-written code. Organizations spend 30% of their IT budgets just managing it. That's not building new things. That's fixing old things that were built wrong.

With AI, this gets worse in a faster way. Google's 2024 DORA [DevOps Research and Assessment] report found something interesting: a 25% increase in AI usage leads to a 7.2% decrease in delivery stability. With the increase in AI-generated code, most developers now spend more time debugging than they did before AI tools existed (Google Cloud, 2024).

Think about what that means for your prototype. You use AI to move faster. You skip documentation because *"we'll add that later."* You don't track prompt changes because *"it's working fine right now."* Three months later, nobody remembers why the AI behaves the way it does, and changing one thing breaks three other things.

> **The speed that made AI prototyping attractive becomes the thing that slows you down.**

Here's What Nobody Tells You About Prototypes

Prototypes rarely stay prototypes.

The most dangerous phase in AI development is the unmanaged transition from "test" to "production." While a prototype is designed to answer a specific question *("Can we do this?")*, a production system must satisfy strict requirements for security, scalability, and maintenance. The problem arises when stakeholders see a functional interface and assume the underlying engineering is complete.

I experienced this firsthand with my Lovable prototype. While the interface worked perfectly after one day of building, the system lacked the security and data persistence required for real users. The distinction between a "quick test" and a "production system" evaporated the moment stakeholders saw it working.

This maintainability gap is critical. Amazon CEO Andy Jassy noted that AI reduced Java upgrade times from six weeks to six hours, saving 4,500 developer-years of work (Jassy, 2024). However, such efficiency is only possible when the underlying code is structured and understood—qualities often missing in rushed prototypes. Therefore, you must assume your prototype will eventually become production code. The goal is to ensure that when this transition happens, the system is clean enough to survive.

The Four Habits That Actually Help

Here is the system I use to move fast without creating chaos: four habits that take less than five minutes each to set up at the very start of a project, and save weeks of cleanup later. It is best to treat these not as optional tips, but as a cohesive routine. Used together, they make the difference between a prototype that helps you learn and one that haunts you later.

Habit 1: Name Everything and Set a Real Deadline

Every prototype needs a name and an end date. Not "the AI thing" or "that conversation test." Something specific like "checkout-assistant-v1" or "support-triage-jan2025."

The name forces you to be clear about what you're testing. The date forces you to actually decide something instead of letting it drift. Here's what this looks like:

Testing a conversational checkout helper:

- **Name**: checkout-assistant-v1
- **End Date**: January 30, 2025
- **Decision point**: If 70% of test users finish checkout without help, we build it properly. Otherwise, we try something different.

The deadline isn't random. It's when you commit to deciding: keep going, start over, or drop it. No *"let's just keep working on it for another month"* without a specific reason.

One thing I learned from product leaders using tools like Lovable: define small, specific tasks instead of uploading entire Product Requirement Documents (PRDs). Describe what users will actually do, step by step. AI works better that way, and so do your teammates when they need to understand what you built.

Habit 2: Keep Your Experiments Separate

When you're prototyping, you write two kinds of code: **disposable code** (quick hacks designed to test an idea) and **keeper code** (the clean, reliable logic you plan to maintain). The disposable code is totally fine. Just don't mix it with the keeper code.

I create a folder called `/experiments` or `/prototypes`. Everything I'm testing goes there. I use clear names like `experiment-conversation-flow` or `test-payment-modal`.

When something actually works and I want to keep it, I move it out of the folder and rebuild it properly. I add error handling. I write the version I'd want to maintain six months from now.

What belongs in experiments:

- Quick UI mockups
- Different ways to phrase prompts
- Test data generators
- Technical spikes to figure out if something's even possible

What should never go there:

- User authentication
- Payment processing
- Database stuff
- Anything touching real user data

I learned this the hard way. My first Lovable prototype mixed real user flows with experimental features. When I tried to remove a test feature, I broke the checkout because everything was connected in weird ways.

In the past, I simply handed requirements to my engineering team, and they handled the version control. I never touched the code. Now that I am the one building with AI, I need my own safety net. I use "isolated branches"—which is just developer-speak for making a duplicate copy of the project before trying something risky. This keeps my experimental mess separate from the version that actually works. If the experiment fails, I delete the copy. If it succeeds, I know exactly what to ask the engineers to build.

Habit 3: Track Your Prompt Changes

Here's something most teams completely ignore: your prompts are as important as your code.

When you change how you prompt the AI, you change how your product behaves. **Think of this like saving different versions of a system's settings or business rules in traditional software.** You need to track these versions so everyone knows exactly what changed and what to expect.

Yet, most people don't do this. Someone tweaks the system message, the AI starts responding differently, and six weeks later, nobody can remember what it used to say or why it changed. Version control for AI prototypes solves this by helping teams track changes, see how designs have evolved, and establish the transparency and accountability needed to revert to previous versions when things break.

I keep a simple file called `prompts.md` in my repo. Every time I update a prompt, I add a version number and write down what changed.

Here's my template:

PROJECT LOG: Checkout Assistant

Version 1.0 (Jan 10, 2025)

- **Goal:** Test if conversational checkout makes sense.

- **Constraint:** Keep responses under 20 words.
- **System Prompt:** "You are a helpful checkout assistant. Guide users through purchasing items clearly and concisely."

Version 1.1 (Jan 15, 2025)

- **Goal:** Fix user confusion on what to say next.
- **Change:** Added explicit options to show choices.
- **System Prompt:** "You are a helpful checkout assistant... If they seem confused, offer to show them a list of options."
- **Result:** Completion rate went from 62% to 74%.

This log becomes valuable when something breaks or when someone asks, *"Why does the AI say it that way?"*

Habit 4: Actually Decide Something

Most prototypes don't fail. They just exist in a zombie state, half-built and half-forgotten, eating up time without delivering value. Making a firm decision is often the hardest habit to master. Immediately after your user testing sessions—where you watch real people interact with your build—you must choose one of three paths:

- **Graduate it**: Build it properly for production
- **Iterate**: Run another experiment with specific changes
- **Shelve it**: Write down what you learned and stop working on it

Product leaders at places like Microsoft and Atlassian follow a similar pattern: start with a one-pager covering the main cases and edge cases, use AI tools to create variations, apply your brand styles and real sample data, then run 5-8 quick user sessions to see what works.

What to Measure

Don't overcomplicate your metrics. I usually pick one "hard" number (did they complete the task?) and one "soft" number (did they feel confident doing it?). If you can't define what success looks like in two numbers, you aren't ready to prototype.

How I decide:

Graduate to production if:

- Hit my success metrics (e.g., >70% completion rate or >4/5 satisfaction score)
- No major technical problems
- Clear path to building it right

Run another iteration if:

- Close to success metrics, but need one specific fix
- User feedback points to something obvious to try
- The approach is solid, but the implementation needs work

Shelve it if:

- Users still don't understand after multiple tries
- Too complicated for the value it provides
- Way off from success metrics with no clear path forward

The key is making this decision on the timeline you set at the beginning. Don't let it drift.

Building a Real Example: A Conversational Career Coach

Let me show you how this actually works. I'm going to build a conversational prototype to help women practice difficult workplace conversations. Things like asking for a raise, addressing someone taking credit for your work, or negotiating scope changes.

I'll walk you through my process using Lovable, showing you exactly what I'm doing and why.

Setting Up

Name: career-coach-v1

Timeline: 10 days

Success criteria: If 7 out of 10 test users complete a practice conversation and say they feel more prepared, I'll build it properly.

I'm using Lovable because it handles conversational interfaces and basic data storage without needing backend setup. This lets me focus on whether the coaching flow actually helps people, not on infrastructure.

Keeping Things Separate

I'll create three prototype versions in separate Lovable projects to test different approaches:

1. **career-coach-v1-directive**: AI gives specific scripts to memorize
2. **career-coach-v1-conversational**: AI role-plays the difficult conversation
3. **career-coach-v1-hybrid**: Combines scripts with practice

Each lives in its own container. When I figure out which one works best, I won't try to merge the others. I'll just archive them and keep iterating on the winner.

Tracking Prompts

I'm going to track every significant prompt change in a shared doc. Here's what I'm starting with:

v1.0 - Initial System Prompt (Day 1):

```
You are a workplace coach helping women prepare for difficult conversations.
```

`Ask what situation they're facing, then guide them through preparation.`

I'll update this as I learn what works. My guess is v1.0 won't work well because it's too vague, but I need to test it to know what to change.

Making the Decision

After 10 days and sessions with real users, I'll look at:

Success metrics:

- Completion rate (target: 70%)
- How prepared users feel (target: 4.0 out of 5)
- Would they use it again?

Technical check:

- Does Lovable handle this smoothly?
- What would need to change for production?
- Any security or privacy concerns?

Then I'll decide: graduate, iterate, or shelve.

If I decide to graduate it to production, I won't ship the prototype code directly. I'll use it as a spec for building the production version right.

How This Case Study Used the System

Notice how the Four Habits guided every step of this project? I didn't just code until it worked. I used **Habit 1** to set a strict 10-day limit. I relied on Habit 2 to keep my three experiments from breaking each other. **Habit 3** ensured I knew exactly why the prompts improved. And **Habit 4** gave me permission to throw away the prototype code but keep the proven logic for the final build.

Your Week-by-Week Plan

Here's how to run a clean prototype in 10 days:

Days 1-2: Setup

- Pick a clear name and end date
- Define what success looks like
- Create your experiment folder or project
- Write your first prompts and version them

Days 3-5: Build and learn

- Get something working
- Keep experimental stuff isolated

- Version every prompt change
- Test with your team

Days 6-8: User sessions

- Run 5-8 sessions with actual users
- Take notes on what works and what doesn't
- Update prompts based on feedback
- Track each version and why you changed it

Days 9-10: Decision time

- Look at your success metrics
- Check technical feasibility
- Make the keep, iterate, or shelve decision
- Document what you learned

If you graduate to production, spend another week rebuilding it properly. Don't rush the production version just because the prototype was fast.

Common Mistakes I've Made

Mistake 1: Thinking In-Memory State Is Production-Ready

Your prototype stores conversation history in memory. That's fine for testing. But when the page refreshes, everything disappears. For prototypes, relying on the browser's temporary memory is fine. However, production apps require dedicated cloud databases (like Firebase or Supabase) to ensure data isn't lost when the user closes the tab.

The fix (See Habit 4): For production, implement proper data persistence from the start. Don't try to add it later.

Mistake 2: Skipping Error Handling

Your prototype assumes everything works. The AI responds correctly. The network doesn't fail. Users don't try weird things. Real production is messier.

The fix (See Habit 4): When graduating to production, add error handling for:

- API failures
- Network timeouts
- Unexpected user inputs
- Rate limiting
- Authentication problems

Mistake 3: Not Writing Down Why You Made Choices

Six months later, someone asks, "Why does the AI respond this way?" and nobody remembers.

The fix (See Habit 3): Keep a decision log with your prompt versions. Note what changed, why you changed it, and what happened as a result.

Mistake 4: Letting Prototypes Linger

The prototype isn't good enough to ship, but not bad enough to kill. So it sits there, consuming attention and time.

The fix (See Habit 1): Honor your end date. Make a decision. If you need more time, fine, but set a new deadline with specific criteria. Don't just *"keep working on it."*

Deciding When to Use This Approach

This rapid prototyping works great for testing ideas, but it's not right for everything.

Don't use rapid prototyping for:

- **Sensitive data systems**: Patient records, financial transactions, and personal information need security from day one
- **Mission-critical features**: If failure causes real harm, skip the prototype

- **Well-understood problems**: If you know the solution works, prototyping just adds steps
- **Complex integrations**: Connecting to multiple existing systems often takes longer than the prototype itself
- **Regulated domains**: Healthcare, finance, etc. require compliance work that makes rapid prototyping impractical

Use rapid prototyping when:

- You're testing a new interaction pattern
- You're not sure users will understand the interface
- You want to validate an AI approach before committing
- You need to show stakeholders something working
- The cost of being wrong is low enough that a quick test makes sense

What Technical Debt Really Means in 2026

The definition of technical debt is shifting. The real problem is that AI magnifies the gap between clean systems and messy ones. Prototype choices compound quickly. One undocumented prompt change makes the next one harder to track, leading to a system that works but nobody understands. This is dangerous because nearly half of AI-generated code snippets contain at least one security flaw (Ficus Technologies, 2024), and even when the code is secure, it often looks correct but lacks internal consistency.

What Success Looks Like

Let me paint two different pictures of the same team six months from now.

Scenario A: Fast and Messy

The team prototyped five features quickly. All five are in production. All five work, mostly. But:

- Nobody remembers why the AI responds the way it does
- Changing one feature breaks another in unexpected ways

- The prompt controlling the AI is 50 lines long, and nobody knows which parts matter
- New features take longer because the codebase is fragile
- The team spends 40% of their time debugging instead of building

Scenario B: Fast and Intentional

Same team, five prototypes. Three graduated to production after being rebuilt properly. Two got shelved after testing. The three in production:

- Have documented prompts with version history
- Were rebuilt with proper error handling and security
- Have clear separation between core logic and experiments
- Can be modified without fear of breaking things
- Keep delivering value without constant maintenance

The only difference? The four habits: naming experiments, isolating code, versioning prompts, and making firm decisions. Same speed. Different outcomes.

Moving Forward

AI prototyping tools keep getting better in 2026. Tools like Lovable, Databutton, and Bolt let product teams add AI to their discovery process. The tools are improving. The risk of accumulating technical debt is also increasing.

The teams that win aren't the ones moving fastest. They're the ones moving fast *and* staying clean. That means:

- Starting every prototype with a clear name and deadline
- Keeping experimental code separate from production code
- Treating prompts like code and versioning them
- Actually deciding what to keep and what to discard

These habits feel like extra work when you start. They feel like insurance three months later when your prototype is still maintainable and your competitors are drowning in technical debt. Your prototype will probably become your production system. Build it like it matters from day one, even when you're moving fast.

Appendix: Templates You Can Copy

Here are three templates I use. Feel free to copy them (see the book website for these and other chapter resources).

Template 1: The Experiment Setup Card

Experiment Name: ________________

- **Timeline:** [Start Date] to [End Date]
- **Purpose:** Test whether [specific hypothesis]
- **Success Criteria:** [Specific, measurable goal]

Decision Framework

- **Graduate if:** [Condition met]
- **Iterate if:** [Partial success condition]
- **Shelve if:** [Failure condition]

Technical Details

- **Tools:** [Tool/framework name]
- **Isolation Strategy:** [Where the experiment code lives]

Template 2: The Prompt Version Log

Feature: [Feature Name] **Version:** [vX.Y]–**Date:** [Date]

The Prompt:

> "[Insert full prompt text here]"

Change Log:

- **What Changed:** [Explain the specific edit]
- **Why:** [Explain the reasoning]

Results:

- **Metric:** [Value]
- **Observation:** [Key qualitative finding]

- **Next Steps:** [What to try next]

Template 3: The Graduation Checklist

Experiment Decision: [Name] **Date:** [Date]–**Duration:** [Time spent]

Part 1: Success Metrics

- [ ] Metric 1: Actual _____ vs Target _____
- [ ] Metric 2: Actual _____ vs Target _____
- [ ] Metric 3: Actual _____ vs Target _____

Part 2: Technical Assessment

- [ ] No major blockers identified
- [ ] Clear path to production implementation
- [ ] Security/privacy reviewed
- [ ] Performance acceptable

Part 3: User Feedback

- [ ] Users understood the interface
- [ ] Users completed primary task successfully
- [ ] Users would use this feature again

Final Decision

- [ ] **Graduate to production** (Proceed to Part 4)
- [ ] **Run another iteration** (Specify changes: ________)
- [ ] **Shelve** (Document learnings and archive)

Part 4: Graduation Plan (If proceeding)

- [ ] Plan proper authentication
- [ ] Plan data persistence
- [ ] Plan error handling
- [ ] Plan monitoring
- [ ] Estimate rebuild timeline

About the Author

Elena Calvillo *writes* ***Prompt-Led Product,*** *where she documents her journey from Product Manager to AI Builder. Her writing emphasizes templates, checklists, and real examples over theory. As a former Software Engineer, Elena focuses on usable playbooks that help professionals ship AI-powered work with confidence at validating ideas and prototyping successfully. She has built and tested conversational prototypes using tools like Lovable and Cursor, sharing transparent case studies from her own experience. Her work reaches 1,700+ subscribers across 89 countries weekly. She is based in Mexico.*

Chapter 23

The Agentic AI Systems

By Lakshmi Veeramani

Intended Audience*: Software engineers and artificial intelligence (AI) practitioners. This chapter assumes basic software or systems thinking and general AI knowledge.*

1. Introduction

From an operational perspective, an AI agent is an intelligent automation system that typically executes a structured decision loop. In practice, this loop can be described through six observable behaviors:

- Understand a user's goal
- Gather or retrieve relevant information
- Break down tasks into steps
- Execute these steps using tools
- Evaluate the results
- Adapt or retry as needed.

Agentic AI is an approach to artificial intelligence where models—most commonly large language models—are combined with memory, tools, and execution logic to create systems that can plan, decide, and act autonomously to achieve defined objectives. Agentic AI addresses the challenge of efficiently performing end-to-end work, rather than isolated tasks, by moving from content generation to reliable execution in real systems.

Machine Learning (ML) primarily learned from historical data to produce predictions and recommendations that helped automate decision-making in systems. These ML models later evolved into broader AI systems that could reason over rules, logic, and data to support more complex problem-solving, going beyond prediction.

In the generative AI era, GenAI applications augment human intelligence by generating new content across different modalities. These models can work with text, images, audio, video, and data, allowing machines to read, summarize, transform, and create information in ways that were traditionally labor-intensive for humans. As a result, the effort required to memorise, process, and reformat information can be significantly reduced, enabling humans to focus on more creative and higher-value work.

Large Language Models (LLMs), a key category within Generative AI, transformed human–computer interaction by introducing the era of copilots: intelligent assistants that support humans in writing, coding, and decision-making. Systems such as GitHub Copilot and ChatGPT accelerated productivity but remained fundamentally passive, responding only when prompted.

Small Language Models (SLMs) operate similarly to LLMs, but are trained on a specific subset of material relevant to the application domain of interest. This enables SLMs to produce more precise outputs with less computational power.

As generative AI systems mature from producing outputs to supporting decision-making, the next evolution is **Agentic AI,** where models (most commonly LLMs and SLMs) are combined with memory, tools, and execution workflows to operate toward defined goals.

Agentic AI is the confluence of three core technologies:

- **Language Models as Reasoning Engines:** LLMs and SLMs provide probabilistic reasoning and planning over natural language and structured context.
- **Tool-Augmented Action**: Agents can call APIs, query databases, or trigger workflows. (Example: calling an API to create a ticket in Jira)
- **Memory-Augmented Intelligence (Retrieval Augmented Generation (RAG) + Persistent Memory)**: Agents store, retrieve, and ground decisions in long-term memory and domain

knowledge, enabling continuity and autonomy over time. Agents remember the context through:

- **Short-term memory** (like conversation history and current task state) and
- **Long-term memory** (like user preferences, prior decisions, learned patterns)

2. The Foundations of Agentic Systems

From a computational perspective, an intelligent agent continuously cycles through:

- perceiving its environment,
- reasoning about it,
- acting upon it, and
- learning from the outcomes.

In Agentic AI, the four steps in this loop are powered by a language model enhanced with memory, tools, and context awareness. Unlike traditional LLM or SLM interactions, agents maintain context and memory across steps and operate in loops rather than single-turn completions.

Core Agent Components

Table 1. Core Components of an Agent

Component	Purpose	Practical Examples (Today)
Model	Acts as the reasoning and decision-making core of the agent. It interprets input, reasons over context and memory, and decides the next action.	• GPT-5.2 / GPT-4o deciding whether to answer directly or invoke a function in an agent built using OpenAI APIs • Claude Sonnet or Opus deciding when to use a tool via structured tool calls in Anthropic agents

Component	Purpose	Practical Examples (Today)
Memory	Stores and retrieves prior context, intermediate decisions, and knowledge to enable multi-step and long-running tasks.	• RAG pipelines using vector databases (Pinecone, OpenSearch) • Persisted agent state across tool calls in OpenAI agents • Claude agent skills retaining intermediate outputs and task context.
Tools	Enable the agent to act on external systems by executing controlled operations.	• Function calling to invoke REST APIs • Database queries • Triggering workflows (CI/CD pipelines, ticket creation, pricing engines) • Claude tool use for querying systems or executing predefined actions
Goal	Defines the objective or task the agent is expected to accomplish and determines completion criteria.	• "Generate an insurance quote from submission data" • "Investigate production errors and summarize root cause" • "Validate a contract against policy rules"
Environment Interface	Connects the agent to users and systems, handling inputs and delivering outputs or actions.	• Chat-based UIs • Backend services sending events to agents via webhooks or queues • Dashboards displaying agent outputs for review and approval

3. Architectures of Agentic Workflows: The Agent Stack

Every agentic system operates on a multi-layered architecture, called the Agentic Stack.

An agent continuously operates in a loop of **perception, reasoning, action, and learning**. While this may sound abstract, each step corresponds to practical system behavior.

3.1 Perception—Understanding What Is Happening

Perception is how the agent receives information from the world. This can include:

- User messages
- Uploaded documents
- API responses
- Sensor or system events

For example, an underwriting agent may perceive:

- A new insurance submission
- Attached risk documents
- External risk scores from APIs

Perception is not limited to human input; it includes any signal that enters the system.

3.2 Reasoning—Deciding What to Do Next

Reasoning is where the agent interprets information and decides its next step.

This may involve:

- Understanding the user's intent
- Breaking a task into smaller steps
- Choosing which tool to use
- Evaluating partial results

For example, an agent might reason:

- "I need additional customer data before I can calculate a quote."
- "This document seems incomplete; I should request clarification."

This reasoning is typically performed by language models, which evaluate context and generate structured decisions.

3.3 Action—Interacting with the Real World

Action is when the agent executes something outside its internal reasoning process, such as:

- Calling an API
- Querying a database
- Sending an email
- Triggering a workflow

For instance:

- Creating a support ticket
- Fetching account balances
- Submitting a pricing request

Without actions, agents remain assistants. With actions, they become participants in business workflows.

3.4 Learning—Improving Behavior Over Time

Learning in most agent systems does not mean retraining the model. Instead, it refers to:

- Storing successful strategies in memory
- Remembering past mistakes
- Adapting future decisions based on outcomes

For example:

- Remembering which data sources were most reliable
- Adjusting routing decisions based on past approvals
- Improving prioritization of tasks

This form of learning improves operational behavior without changing the underlying model.

Table 2. Agentic Architecture Layers

Layer	Function	Example Tools / Frameworks
Model Layer	Core reasoning and language understanding	GPT, Claude, Mistral, Llama, Gemini
Runtime Layer	**Agent** orchestration, execution, and protocol coordination	LangChain, LlamaIndex, AutoGen, MCP (Model Context Protocol)
Memory Layer	Context retention and retrieval across reasoning cycles	Pinecone, Weaviate, Redis, Open Search
Action Layer	Tool usage and environment interaction	API calls, plugins, microservices
Governance & Observability Layer	Policies, safety, telemetry, and lifecycle management	OpenTelemetry, Arize, Guardrails, AgentOps
Interface Layer	Human interaction, visualization, and control panels	Streamlit, React, Dashboards

This architecture ensures modularity and scalability, so that agents can grow in complexity without losing control or transparency.

4. Language Models in Agents: LLMs and SLMs

Different tasks place different demands on language models, which is why agentic systems often combine models with varying capabilities. This approach helps balance performance, cost, and latency while maintaining effective reasoning.

Table 3. Characteristics of LLMs and SLMs

Characteristic	LLM (Large Language Model)	SLM (Small Language Model)
Power	High reasoning and creativity	Task-specific intelligence
Latency	Higher	Lower
Deployment	Cloud-based	Edge / on-prem
Use Case	Complex multi-step reasoning	Lightweight, embedded inference

In practice, hybrid architectures combine both:

- An LLM for deep reasoning and orchestration.
- Multiple SLMs for focused sub-tasks (data extraction, sentiment analysis, etc.).

In practice, LLMs are used for complex tasks such as document analysis, multi-step reasoning, and decision orchestration, while SLMs handle focused tasks like classification, validation, routing, or data extraction.

In a hybrid setup, SLMs often act as fast, low-cost filters or controllers that decide when a request needs deeper reasoning from an LLM, reducing overall cost and latency. LLMs are typically cloud-based due to their size and compute needs, while SLMs run efficiently on edge or on-prem systems, offering better scalability and lower cost, though with more limited accuracy and reasoning depth compared to large models.

5. Tools and Skills in Agentic Systems

Unlike LLMs, agents not only think; they act. They need "tools" for the action part.

Tool-Augmented Action

Agents can call APIs, query databases, or trigger workflows. Tools extend an agent's cognitive reasoning into real-world execution.

Common Tool Types

Reasoning becomes valuable when agents can translate decisions into actions in real systems. Tools provide this connection by allowing agents to interact with APIs, databases, workflows, and external services. Together, these tools define what an agent can practically accomplish beyond language generation.

Table 4. Common Tool Types Used in Agentic AI Systems

Tool Type	Example	Purpose
API Connectors	REST, GraphQL	External integrations
Code Interpreters	Python, JS	Execute computations
Knowledge Tools	RAG APIs	Fetch context
Actuators	Webhooks, IoT controls	Real-world actions
Data Interfaces	SQL, BigQuery	Structured data operations

The recent "skills" introduced by Claude are indications that the Agent concept is still evolving and the Action part is getting more augmented (Anthropic, 2025).

6. Memory in Agents

Memory enables agents to persist context beyond a single interaction. Memory is a key capability that allows agents to operate across multiple steps and interactions. By retaining context, prior decisions, and retrieved knowledge, agents can maintain continuity and make informed decisions over time.

Table 5. Types of Memory in Agentic AI Systems

Type	Function	Implementation
Short-Term Memory	Retains recent context during conversation	Context windows, buffers
Long-Term Memory	Stores embeddings and historical results	Vector databases (Pinecone, FAISS)
Episodic Memory	Captures experience over time	Temporal embeddings
Semantic Memory	Encodes knowledge structures	Knowledge graphs
Working Memory	Active reasoning buffer	LLM token window

Memory Management Strategies

For agents to work reliably over time, memory must be stored, retrieved, and reused in a structured way. Four common strategies address how raw information is converted into usable context that supports reasoning and decision-making.

1. **Chunking**: Split large documents into retrievable units.
2. **Embedding**: Convert chunks into semantic vectors.
3. **Retrieval**: Query memory using contextual similarity.
4. **Reinforcement**: Store outcomes of successful actions.

When developing the Agentic AI systems, these steps must be carefully thought and suitable strategies applied. For example, the chunking strategies for Excel and PDF files will be different.

7. Agentic Workflow Patterns

Agents execute tasks through structured loops and coordination models. Agents operate using well-defined execution patterns that shape how they respond to inputs, coordinate tasks, and refine outcomes. These patterns reflect different design choices depending on the level of autonomy and control required.

Common patterns include:

1. **Reactive Loop** : Acts in response to user input.
2. **Proactive Loop**: Observes environment changes and acts autonomously.
3. **Collaborative Loop**: Multiple agents coordinate for a shared outcome.
4. **Hierarchical Loop**: Supervisor agents manage sub-agents.
5. **Reflective Loop** : Agents analyze their reasoning and self-improve.

Each workflow reflects a design philosophy: reactivity for efficiency, proactivity for autonomy, and reflection for safety.

Refer to (Veeramani, 2024; Veeramani, 2025) for more details on the workflow patterns.

8. Agent Development Lifecycle Activities

Building an agent requires more than prompt engineering. It involves a disciplined development lifecycle that treats agents as engineered systems rather than experimental prototypes.

Table 6. Stages of the Development Lifecycle for Agentic AI Systems

Stage	Description
Plan	Define the user's goal, break it into tasks, specify success criteria, identify risks, and determine required tools.
Data (Context & Retrieval)	Use retrieval systems, summarization, and compression to construct high-quality context. Track provenance to ensure trust.

Stage	Description
LLM (Model Fleet Selection)	Use multiple models: small models for routing and validation, medium models for planning, and large models for synthesis. Evaluate models using cost, latency, accuracy, robustness, and consistency.
Prompt	Prompts shape agent behavior. Use roles, reasoning scaffolds, schema constraints, verifier prompts, and reflection prompts.
Tools	Agents rely on tools such as APIs, search, databases, and code interpreters. Use Model Context Protocol (MCP) for tool discovery, schema definitions, permissions, and versioning.
Code (Agent Implementation)	Implement the agent runtime, orchestration logic, state management, retries, and error handling that coordinate models, tools, and memory during execution.
Output	Outputs must be validated, structured, accurate, and explainable. Use JSON/YAML schemas, confidence scores, and provenance metadata.

9. The Interface Layer - Humans In The Loop

Purpose

Even the most autonomous agents require human oversight. Human-In-The-Loop (HITL) interfaces provide visibility, control, and feedback mechanisms that allow humans to supervise, guide, and intervene when required.

Table 7. Typical Human-In-The-Loop Interfaces

Interface	Function	Example
Dashboards	Monitor agent health	Grafana, Streamlit
Agent Chat UIs	Interactive dialogues	React, Chat Components

Interface	Function	Example
Explainability Views	Display reasoning traces	Arize, Langfuse
Control Panels	Modify policies, restart agents	AgentOps Admin Console

Human-In-The-Loop Design

Key design patterns:

- Intervene on error : allow humans to override actions.
- Approve before execution : review critical tasks.
- Feedback to learning : store human corrections in memory.

10. Multi-Agent Systems

Multi-Agent systems, such as in LangChain, break down complex workflows into coordinated teams of specialized agents (LangChain, 2026). This enables distributed development, parallel execution, and more effective context management than a single monolithic agent. Rather than overwhelming one agent with all tools and knowledge, multi-agent patterns allow each agent to focus on a subset of responsibilities, improving clarity and performance when solving tasks that span domains or require extensive context.

Key patterns include:

- subagents (where a main agent orchestrates subordinate agents),
- handoffs (dynamic transitions between agent behaviors based on state),
- skills (on-demand loading of specialized expertise),
- routers (classification and routing to domain-specific agents), and
- custom workflows (bespoke orchestrations using LangGraph).

At the heart of this design is context engineering: carefully deciding what information each agent should see to maximize effectiveness and reduce unnecessary model overhead, especially when tools and domains grow large.

11. Observability in Agentic Systems

Why Observability Matters

Agents are built using LLMs, which are highly non-deterministic. So when we want to replace business processes which were performed using traditional software, we have to monitor closely for potential misbehavior from this non-determinism. This requires enhancing the observability of every step the agent is performing.

Traditional observability tracks system health. Semantic observability tracks cognitive health, reasoning depth, coherence, bias, and accuracy. Effective management of Agentic AI systems requires both. In simple terms, traditional observability answers *"Did the system run correctly?"*, while semantic observability answers *"Did the system think correctly?"*

For example, if an agent flags an insurance claim as suspicious, classical observability can confirm that the API call succeeded. Semantic observability, however, examines *why* the agent made that decision: what context it retrieved, what reasoning steps it followed, and whether the conclusion was logically consistent.

When an Agent is capable of acting, we need to monitor its thinking to ensure safe operation. Monitoring agent reasoning in production systems typically involves:

- Capturing structured prompt and response logs
- Recording tool invocation sequences
- Storing intermediate reasoning outputs (when allowed by policy)
- Tracking confidence scores or self-evaluations
- Validating outputs using secondary verifier models

For example, an agent may generate a draft decision along with a confidence score. If the confidence is below a threshold (e.g., 0.75), the system can automatically escalate to human review.

The ultimate success of observability for the Agents deployed in production systems is when we are capable of stopping the Agent's action if needed, based on observing its reasoning process. Seeing the 'mind' of the machine is the key here.

Agent thinking → **OBSERVE** → Action

Table 8. Classical vs. Semantic Observability

Dimension of Observability	Classical (Traditional Software)	Semantic (Agentic AI)
Trace	Function calls	Thought chains
Metrics	Latency, uptime	Confidence, accuracy, reasoning depth
Logs	System events	Prompts, responses, reflections
State	Static variables	Dynamic memory embeddings
Root Cause	Stack trace	Reasoning graph

A reasoning graph represents the structured flow of an agent's decision process. Instead of showing function calls like a traditional stack trace, it visualizes how the agent moved from input → retrieved memory → intermediate reasoning steps → tool calls → final output.

For example, a reasoning graph for a claims review agent might show: (1) claim received, (2) prior claims retrieved, (3) risk score calculated, (4) anomaly rule triggered, (5) fraud flag issued. This allows teams to trace not just what executed, but how the conclusion was formed.

Human-in-the-Loop systems, which we discussed earlier, rely heavily on semantic observability. Observability provides the reasoning traces that humans review before approving actions. Without visibility into how an agent reached a conclusion, human oversight becomes superficial. In production environments, semantic traces often feed directly into approval dashboards where operators can validate reasoning before execution.

Some tools supporting semantic observability are:

- OpenTelemetry – Captures distributed traces, logs, and metrics across services, forming the foundation for agent telemetry.

- Arize AI – Provides model performance monitoring and evaluation metrics.
- Langfuse – Tracks prompts, responses, reasoning traces, and user feedback.
- LangSmith – Offers detailed tracing of agent execution graphs and tool calls.
- AgentOps – Focuses specifically on monitoring agent behaviors, cost, and reliability.

12. AgentOps: Managing the Agent Operations

DevOps is the discipline of building, deploying, monitoring, and continuously improving software systems in production environments. As agents move into production, they require an operational discipline similar to DevOps, but tailored to reasoning systems.

AgentOps is to Agentic AI what DevOps is to Software Engineering: a discipline for deploying, monitoring, and governing agents in production environments.

Its key pillars include:

1. **Observability**: Capturing semantic traces.
2. **Governance**: Applying ethical and operational policies.
3. **Reliability**: Ensuring stable, repeatable outcomes.
4. **Optimization**: Managing performance and cost.

Table 9. Stages of AgentOps

Stage	Description	Goal of the stage
Design	Define goals, constraints, and interfaces.	**Ensure the agent is solving the correct problem within safe operational boundaries.** The goal is to formalize objectives, define measurable success criteria, constrain tool permissions, establish escalation rules, and prevent goal drift before deployment.

Stage	Description	Goal of the stage
Deploy	Package and host in scalable environments.	**Achieve reliable, reproducible, and secure execution in production environments.** The goal is to version models, prompts, and tool schemas, isolate environments, enforce access controls and ensure scalability without compromising consistency.
Monitor	Track metrics, logs, and reasoning traces.	**Detect deviations in behavior before they cause business impact.** The goal is to continuously observe infrastructure health and semantic performance (hallucination rate, tool accuracy, confidence levels, escalation frequency) to identify drift, instability, or misuse.
Optimize	Improve reasoning quality and performance.	**Continuously improve accuracy, efficiency, and cost-effectiveness.** The goal is to refine prompts, adjust model routing, tune memory retrieval, reduce unnecessary tool calls, and enhance reasoning reliability through structured iteration.
Govern	Ensure compliance and trustworthiness.	**Maintain bounded autonomy aligned with policy, ethics, and regulation.** The goal is to enforce approval workflows, audit trails, data protection controls, bias mitigation, and clear accountability for agent decisions.

This iterative lifecycle formalizes AgentOps as a continuous improvement discipline.

Mini Case Study: Claims Review Agent in Production

Consider an insurance company deploying a Claims Review Agent.

- **Design:** The goal is to flag suspicious claims above a certain threshold while restricting the agent from directly rejecting claims.

- **Deploy:** The agent is containerized, connected to policy databases, and configured with tool permissions limited to read-only access.
- **Monitor:** Observability tracks reasoning traces, tool calls, and false positive rates.
- **Optimize:** Prompt tuning reduces over-flagging and improves explanation clarity.
- **Govern:** Human approval is required before any fraud escalation is initiated.

Without AgentOps, such a system might:

- Flag too aggressively
- Miss edge cases
- Provide untraceable decisions
- Expose sensitive data

With AgentOps, the system becomes controlled, auditable, and production-ready.

13. Governance and Responsible Agent Design

Autonomy without oversight can lead to bias, misinformation, or unintended consequences. Therefore, responsible agent design must embed principles of fairness, accountability, and transparency.

Explainability is a regulatory and trust necessity to adopt AI Agents into production. By linking each decision to its semantic reasoning trace, developers and users can reconstruct the path of thought, ensuring both accountability and comprehension.

Core Governance Controls

1. Policy Enforcement: Define allowed actions and data boundaries.
2. Bias Mitigation : Audit for representational fairness.
3. Transparency Logs: Record reasoning and outcomes.
4. Explainability Interfaces : Provide "why" alongside "what."
5. Human-in-the-Loop : Enable oversight for critical decisions.

Table 10. Core Components of Governance Controls

Component	Function	Practical Examples (Today)
Policy Engine	Defines what agents can and cannot do by enforcing operational, ethical, and business rules.	Allowing an agent to read customer data but blocking write access Restricting which APIs an agent can invoke Enforcing approval-before-execution for high-risk actions in enterprise agent platforms.
Telemetry System	Captures traces, logs, metrics, and semantic signals for observability and analysis.	LangSmith or Langfuse capturing agent execution traces Logging prompts, tool calls, and outcomes Tracking latency, cost, and reasoning depth for OpenAI or Claude-based agents.
Redaction Layer	Prevents exposure of PII or sensitive data in prompts, memory, logs, and outputs.	Automatically masking names, emails, policy numbers before storing traces Redacting sensitive fields before sending context to LLMs Enforcing tenant-level data isolation in multi-tenant agent platforms.
Audit Store	Maintains immutable records of agent decisions, actions, and outcomes for compliance and investigations.	Write-once audit logs for regulated industries (insurance, banking) Storing agent reasoning summaries and tool usage history for post-incident analysis and regulatory review.
Safety Filters	Validates agent inputs and outputs against predefined guardrails and safety policies.	Blocking hallucinated recommendations Preventing unsafe instructions Enforcing content policies using rule-based filters, classifiers, or model-based safety checks before responses are returned.

Governance components do not operate independently; they function as a coordinated control system. The policy engine defines what actions are permitted. The safety filters validate inputs and outputs before execution. The telemetry system captures reasoning traces and tool usage events. The audit store persists these records immutably for compliance and investigation. For example, when an agent attempts to invoke a high-risk API, the policy engine evaluates permissions, the safety layer checks for rule violations, the telemetry system records the decision context, and the audit store preserves the outcome for regulatory review. Together, these layers ensure that monitoring is not passive observation but active control.

In real-world deployments, safety filters have prevented large-scale misinformation and policy violations. For example, content-generation agents deployed in enterprise environments often include classifier-based safety checks that block outputs containing sensitive data or hallucinated legal advice. Similarly, financial services firms deploy redaction layers to automatically mask customer identifiers before prompts are sent to language models, preventing accidental data leakage. These mechanisms act as runtime safeguards, catching issues before they reach users or external systems.

Human oversight must be calibrated carefully. Too little oversight can allow errors to propagate; too much oversight can eliminate the efficiency benefits of autonomy. A common design pattern is risk-based escalation.

- Low-risk actions (e.g., drafting summaries) may execute automatically
- Medium-risk actions (e.g., pricing recommendations) require human approval
- High-risk actions (e.g., regulatory filings, irreversible financial transactions) always require explicit human authorization.

This tiered approach preserves productivity while maintaining control.

Non-deterministic agentic workflows must operate within deterministic business process boundaries to ensure reliability and control.

Governance requirements vary by industry. In regulated sectors such as banking and insurance, immutable audit stores and explainability traces are mandatory for compliance reviews. In healthcare, privacy controls and redaction layers must comply with data protection regulations such as HIPAA. In autonomous systems (e.g., vehicles or robotics), real-time safety filters must operate with strict latency guarantees. In technology and SaaS environments, tenant isolation and prompt redaction prevent cross-customer data exposure.

While governance mechanisms such as safety filters and policy engines apply universally, their strictness, latency tolerance, and audit depth depend on the risk profile of the domain.

14. Designing Scalable Agents

Agentic AI systems require cloud-native architectures to scale reliably, remain observable, and operate securely in production environments. Cloud-native design allows agents to be deployed as modular services, coordinated through events, and governed with strong operational controls.

Unlike experimental prototypes, production-grade agents must support asynchronous execution, fault tolerance, and secure tool access. This requires architectural discipline similar to modern distributed systems.

Table 11. Core Components of Scalable Agent Design

Component	Function	AWS-Specific Examples (Today)
Containerized Agents	Package agents as isolated, deployable services for scalability and modular updates.	Dockerized agents deployed on Amazon ECS, Amazon EKS, or AWS Fargate; each agent running as an independent microservice with its own lifecycle.

Component	Function	AWS-Specific Examples (Today)
Event Streams	Coordinate agent execution, handoffs, and state transitions using events.	Amazon EventBridge for event-driven orchestration; AWS Step Functions for long-running agent workflows; Amazon MSK (Managed Kafka) for high-throughput event streams.
Telemetry Pipelines	Capture traces, logs, metrics, and semantic signals for observability.	AWS Distro for OpenTelemetry (ADOT) collecting agent traces; logs stored in Amazon CloudWatch Logs; metrics published to CloudWatch Metrics and exported to observability tools.
Vector Memory Stores	Store and retrieve embeddings for semantic memory and RAG.	Amazon OpenSearch Service with vector search enabled; Amazon Aurora + pgvector for structured vector storage; Amazon Bedrock Knowledge Bases for managed RAG pipelines.
Auth Layers	Secure agent access to tools, APIs, and data using identity and access controls.	AWS IAM roles for agent permissions; Amazon Cognito for user authentication; OAuth/JWT integration for secure API access; fine-grained IAM policies controlling tool invocation.

Key principles for cloud-native agentic systems include:

1. Containerization for Portability

Purpose: Ensures consistent execution across environments while enabling controlled updates and independent scaling.

Example: A claims review agent deployed as a container can be promoted from staging to production with identical model and tool configurations.

2. Event-Driven Workflows for Responsiveness

Purpose: Allow agents to process tasks asynchronously, decoupling reasoning cycles from upstream systems.

Example: A transaction event triggers a fraud detection agent, which publishes a risk decision without blocking transaction processing.

3. Telemetry-First Design for Observability

Purpose: Embed infrastructure and semantic monitoring into the architecture from the beginning.

Example: A decline in reasoning confidence from an underwriting agent triggers alerts before incorrect outputs scale.

4. Immutable Logs for Compliance

Purpose: Preserve prompts, tool calls, and outputs in tamper-resistant storage for audit and regulatory review.

Example: During an investigation, an organization reconstructs the full reasoning path of a loan approval agent.

5. Scalable Vector Storage for Memory Recall

Purpose: Enable efficient semantic retrieval as long-term memory grows.

Example: A customer support agent retrieves relevant historical embeddings without exceeding token limits.

15. Conclusion: The Discipline of Agentic AI

If there is one thing I want to convey through this chapter on Agents, it is this:

Agentic AI systems are powerful not merely because they use large language models, but because they combine probabilistic reasoning with structured memory, controlled tool access, deterministic

workflows, and governance mechanisms that make autonomy reliable.

The shift from traditional software to agentic systems represents a movement from static logic to adaptive reasoning. However, adaptive systems must be architected with discipline. Without boundaries, observability, and operational controls, autonomy becomes risk. With them, autonomy becomes leverage.

Consumer-facing systems such as ChatGPT by OpenAI, Claude by Anthropic, and platforms like Perplexity AI illustrate early agentic capabilities—particularly when features such as tool use, custom GPTs, workflow automation, or code execution are enabled. These platforms can serve as practical entry points for experimenting with tool-augmented reasoning. However, enterprise-grade agentic systems extend far beyond conversational interfaces. They require memory management, policy enforcement, auditability, telemetry pipelines, deterministic process boundaries, and structured operational governance—the architectural components discussed throughout this chapter.

If you work in technology, architecture, or operations, understanding how language models can orchestrate workflows in your domain is becoming essential. The question is no longer whether agents will be deployed, but how responsibly and effectively they will be engineered.

The promise of Agentic AI is not machines replacing humans, but machines thinking with humans: at scale, with transparency, and with purpose.

About the Author

Lakshmi Veeramani *is helping organizations unlock the full potential of Agentic AI—from strategy to scalable implementation. Her focus is on building trustworthy, production-ready Agentic AI systems through AgentOps, observability, and governance frameworks that ensure long-term success. With expertise in AI/ML, data science, and AgentOps, she builds trustworthy and governed AI agentic systems. A recognized thought leader,* *she shares insights on AI strategy, enterprise AI adoption, guiding organizations to move beyond pilots toward outcome-driven AI innovation, through Substack and LinkedIn. Based in India.*

Chapter 24

Stop Prompt Gambling

How to Govern AI-Human Collaboration

By Iwette Rapoport

Intended Audience: *This chapter is for professionals using large language model (LLM) chatbots in real work—where speed, credibility, and accountability matter. You'll learn a practical way to govern a conversation: how to set roles, slow down drift, and keep evidence ahead of fluent but unverified output.*

MY EARLY WORK with generative AI began in 2021, before the public knew the name ChatGPT. GPT 2.5 was very early and unsophisticated, and it showed. We were teaching a difficult domain (domestic abuse) and how to reroute conversations safely. When doing this, we hit real friction. In one test, I wrote that I was afraid of my partner. The model replied, "Are you afraid of me?" and then looped. By today's standards, GPT-2.5 was extremely limited compared to modern LLMs.

Being so early clarified my approach: when working with LLMs, you need to interact in a way that creates depth and mutual learning. Only then can you get the system's full value. To get there, you have to invest time, agree on shared definitions, and set reciprocal protocols: how you prompt, how the system responds, how you verify, and how you correct. This is what I mean by governing your thread. No institution has more influence over a given chat thread than the people inside it.

Today's Systems

Today's generative AI models can plan multiple steps, use tools, and work across text, code, images, and audio. When you connect them to your own material, they can answer from your documents rather than guessing. When you let them use tools (such as code execution, document search, image generation, or automation workflows), they can do real work: advanced calculations, drafting assets, testing alternatives, or assembling parts of an application before returning a single best answer. They can also try several ideas at once and then give you one best answer. Another fact is that the same prompt rarely yields the same reply. This is because the model is generating a best next response each time based on probabilities – not replaying a fixed script – so small differences in context or randomness can change the outcome.

Example: Context & mode dependence

I once copied a prompt that looked excellent: clear instructions, well-phrased, and tested by someone else. When I pasted it into a new thread, the output was flat and generic. Nothing was "wrong" with the prompt itself. What was missing was the surrounding context: earlier definitions, prior decisions, and shared assumptions that conditioned how the instruction was interpreted.

This is why governance must live in the chat thread. The same instruction means different things depending on what came before it. Therefore, the same prompt can generate different results for person A compared to person B; each output reflects prior interaction.

From GPT 4.0 onward, you can ask the model to follow a small plan or call functions; it can handle multi-step work. Bottom line: these systems are fast, fluent collaborators. They get even better when we show them where facts come from, mark what's an assumption, and keep a simple record of decisions.

If you want collaboration and the best output, I urge you to stop **prompt gambling** and instead start **role-based collaboration**. Collaboration/interaction include roles, iterations, and shared rules inside a chat thread. You can even direct the model's reasoning, not just its wording. For example, a collaborative thread can set roles explicitly: the system provides analysis and information; the human provides context, checks the analysis, and makes the decision.

Why governance must live in the thread

The deeper you collaborate, the more you notice what isn't visible.

Even with clear roles and iteration, there comes a point where the model's reasoning vanishes into its own architecture. Modern systems layer weights, fine-tuning, and tools in ways that are rarely traceable end-to-end. Even engineers can usually inspect parts, not the whole. In practice, transparency collapses into partial views. What remains reliable is traceability at use. That is why governance shifts from seeing inside the model, waiting for policies to form, to seeing inside the interaction:

- what data inputs entered,
- what claim emerged,
- what reasoning path was visible, and
- what we chose to accept.

That is how both we and the system learn what is acceptable output in our interaction.

Governance-in-Thread

Governance-in-thread means the conversation itself is the operating surface:every claim, number, and decision meets small rules that keep evidence ahead of smooth-sounding but unverified output. I use this phrase to mean governance applied inside the conversation (the thread): small rules that slow down claims until evidence, scope, and ownership are explicit. The thread is the unit of governance: the conversation itself is where claims are made, checked, and either accepted or revised. If a draft says "turnover dropped," we treat that sentence as *provisional* and ask: **Which metric? Over what period? Compared to what baseline? Source link?** Only then do we keep the sentence, or rewrite it to match the evidence. Governance-in-thread turns quality control into continuous behaviour rather than after-the-fact review. The endpoint is co-regulation: either the system or the human can slow down, question an assumption, or label uncertainty. A quick friction check can flip the mode from "Creative Build" to "Public-Facing"; when it flips, the writing tightens and adds sources. Over time, this builds learned restraint in the system. The system begins to anticipate governance as part of dialogue, not an interruption.

I have found it best when both sides co-stabilise the interaction. One of the protocols that is applied here is the friction check. For example: One day we were working in business mode and I drifted into something else, and wrote a sentence. The system responded: *"Excellent—let's continue with the agreed task."* That response acted as a refocusing cue: it reinforced the shared objective and pulled the thread back to the scope we had set. In other words, the system supported the protocol rather than just generating content.

To make governance-in-thread operational, we keep micro-protocols: Fact Gate (our internal check: quantitative or causal claims must be sourced or tagged [Missing Source] before publication), Mode Labels, Friction Check, Authorship Log). If a statistic lacks a source, we tag **[Missing Source]**, switch to Public-Facing mode, find the study, and link it back. We protect against a common failure mode. Under pressure, we pause and we rewrite absolute claims into precise ones. After a session, we can export an excerpt with claims, owners, and links as an appendix. The audit trail is born from the conversation itself.

Example: Fact Gate

In a draft, the system wrote: *"Employee turnover dropped significantly after the intervention."* The sentence sounded right, but no metric or source was visible. Instead of rewriting it to sound better, we paused the sentence and applied the Fact Gate. We tagged it **[Missing Source]** and asked: *Which turnover measure? Over what time period? Compared to what baseline?*

Until those questions were answered, the sentence stayed quarantined. Once the source was found and linked, the sentence was corrected.

Governance-in-thread turns fact-checking into a habit, not a final review.

True meaning of co-regulation

Co-regulation sounds balanced and easy, but in practice it is demanding for both human and system. It depends on explicit structure: what each side can and cannot do. When it fails, the thread can collapse into performance: the human improvises control, the system mirrors confidence, and judgement dissolves into fluency—prioritising smooth, confident-sounding text over accuracy.

Co-regulation is best understood as a disciplined 5-part architecture of roles and rules.

A. **Permission:** Co-regulation only works if **roles are explicit**. Each participant knows who carries authority, who carries analysis, and who carries veto power, which remains with the human. The thread must name who can halt motion. Without a defined stop rule, speed becomes the governing force.

B. **Pause:** It only works if **pauses are protected.** Friction is treated as information. When either side signals unease, the pace slows without the system being conditioned to push through. If not, the system may pattern toward overriding uncertainty, and the human may learn to suppress intuition.

C. **Proof:** It only works if **evidence precedes fluency.** No claim should stand because it sounds right. Each elegant paragraph is guilty until proven factual. The habit of tagging **[Missing Source]** or calling a **Fact Gate** turns opinion into verifiable sequence.

D. **Drift Detection:** It only works if **drift is detectable.** Semantic drift—when meaning slides, while words remain stable—is a known failure mode in language systems. Each session needs a stabilisation check: have our key terms, risk thresholds, or moral framing shifted since the start? If yes, restate the agreed definitions and re-anchor before continuing.

E. **Asymmetry:** And it only works if **asymmetry is acknowledged.** The human holds accountability; the system holds assistance. When either forgets, the loop inverts. In the worst case, the model begins to decide, or the human begins to mimic. The contract must be restated whenever the asymmetry blurs.

Co-regulation, then, is not a harmony, but a rhythm of governed tension. It keeps both sides awake. When it works, the conversation becomes clearer.

From principle to practice

Our work happens where rules live: every claim, number, and decision meets small gates that keep evidence ahead of fluency. Core micro-protocols include Fact-Gate, Mode Labels, Friction Check, and protection from echo chambers. Some of these protocols are equally useful in human teams. For example, a **Fact Gate** interrupts groupthink by requiring sources before confident claims are accepted, reducing the risk of consensus forming around unverified statements. A **Friction Check** can counter "not invented here" dynamics by pausing momentum and asking whether resistance is based on evidence, ownership, or identity rather than merit. In both cases, the goal is not to slow work unnecessarily, but to prevent speed and social pressure from becoming the default decision-maker.

Rules we keep

We treat the conversation as the operating surface: every claim, number, and decision meets small rules that keep evidence ahead of fluency. Our work happens where rules live.

Fact Gate and Mode Labels

Rule: Fact Gate, Mode Labels, Authorship Log.

The Fact Gate is the simplest control in our governance-in-thread practice: no claim goes public without a visible source, a clear [Missing Source] tag, or an explicit [Personal Opinion] label. This turns fact-checking into a structural reflex rather than a late-stage edit. When a statement enters the public-facing zone, we run it through the gate: verify, cite, or tag. Unverified lines remain quarantined until evidence arrives.

Mode Labels define the context in which reasoning occurs (for example, *business* or *creative* mode). Each label carries its own discipline: exploratory looseness in creative mode, analytical restraint in business mode. Switching labels mid-thread signals a shift in responsibility and tone.

Together, Fact Gate and Mode Labels generate an Authorship Log. They allow us to trace not only *what* was said, but *under which mode* it was said, *by whom*, and *whether the claim was ever verified.*

Friction Check

Rule: Pauses protect judgment.

Friction is information. When a thread slows, something valuable surfaces. In mechanical systems, we use heat sensors to detect strain; in collaborative systems, friction plays the same role. It marks the edge where speed starts eroding judgment.

A friction protocol doesn't punish delay; it formalises the pause. In practice, this means:

- Naming the moment of unease instead of smoothing it over.
- Switching from output to reflection mode. *"Why did this feel off?"*
- Recording what triggered the tension: missing evidence, unclear intent, ethical risk, fluency, drift.

Each pause is a small act of care. Over time, those moments can *teach you where integrity tends to slip:* the situations where speed outruns judgment or where unverified claims feel "good enough." You don't need a formal log for this to work; even noticing the pattern is governance.

In high-velocity environments, absence of friction is not flow; it's drift. Friction reminds us that good systems are not the ones that never slow down, but the ones that know why they do.

Example: Friction Check Protocol

While drafting a governance brief for a 30-day pilot under **GATE at Change Collective**, the system and I began agreeing too easily. Each revision improved the phrasing, but the brief was drifting: smoother claims, no added precision. The thread had shifted into momentum. It felt fast ... and wrong.

I called a Friction Check and stopped. We rolled back to the last verified step (reason visible; sources linked or tagged [Missing Source]), marked the rest as drift, and restated roles: the system analyses; I decide. Only then did we continue. The point was to exit safely without losing traceability, before speed erased judgment.

Purpose: Maintain coherence under acceleration. **When to trigger:** Whenever a claim feels too fluent, a decision too easy, or a result too certain.

Steps:

1. **Signal**—type Friction Check.
2. **Switch mode**—move from *Creative Build* to *Reflective* mode; slow output to half speed.
3. **Locate**—identify the element causing tension (unsupported claim, drift).
4. **Log**—record a short note: trigger, action, follow-up owner/date.
5. **Return**—resume build once evidence, intent, or boundary is restored.

The protocol creates a visible hinge between spontaneity and accountability. It lets a collaboration breathe without losing traceability. What begins as a momentary slowdown becomes a muscle memory for integrity.

Tip: Protection from echo chamber collaboration prompts

One way to enhance the effectiveness of your Friction Check is to flip the causal: "If our conclusion were false, what would we expect to see? Do we see it?" Another way is to define falsifiers: "List 3 observations that would invalidate our claim."

Template to use

Every collaboration needs a working agreement, a short declaration that defines who decides, who analyses, and how truth is kept alive between them.

- **Human role:** authority and final judgment. The human defines purpose, interprets output, and decides when a result is safe to use.
- **System role:** analysis, sourcing, and draft generation. The system extends reach, not responsibility.
- **Evidence rule:** every claim must trace to a source or carry a [Missing Source] tag until verified. **[Personal Opinion]** marks value judgments so they don't masquerade as facts.
- **Friction rule:** either side may call pause; the human decides when to resume.
- **Archive rule:** every major decision, citation, and revision is logged (human-initiated, system-assisted).

In one recent business-case session, the system produced a cost-of-inaction estimate for turnover related to mental ill health related turnover. It flagged [Missing Source] for one assumption. I located the Deloitte 2024 report, confirmed the figure, and re-ran the brief. The audit log captured both entries, and the thread advanced with evidence.

From Thread to Organisational Governance

What begins as discipline in a single conversation can scale into organisational integrity. When micro-protocols such as fact gates, friction checks, and authorship logs are embedded into daily collaboration tools, governance turns into literacy rather than oversight. The intranet becomes a shared operating field where small behaviours like tagging [Missing Source], marking pauses which teach both human and AI system verification as a habit.

These practices can later transfer to purely human teams. Each team thread becomes a unit of integrity. Managers learn to co-govern their decisions in real time, not through post-hoc reports. The organisation gains a living record of how conclusions were reached, what was questioned, and when evidence entered the process.

After each session, exporting the thread excerpt (claims, owners, and links) creates an immediate record. Over time, these excerpts form a distributed audit trail: a history of reasoning born directly from the work itself. Governance, then, is no longer bureaucracy; it becomes traceable conversation.

Failure Modes and Safe Exits

Co-regulation is designed to hold balance, but every collaborative system faces moments when it slips. The most common failure modes appear:

- when speed overrides depth,
- when fluency replaces verification loops, or
- when the model begins to mirror emotional tone instead of analysing it.

These moments are evidence of human–system feedback loops running too fast for judgment to keep pace. When co-regulation slips, both sides begin reinforcing each other's last output instead of introducing new reasoning. The human responds to the system's previous line; the system patterns on their phrasing; the loop amplifies itself. The exchange accelerates faster than either of the ones in the thread can pause to check why it's moving that way. Words keep flowing, but the reasoning no longer advances. In that moment, something shifts in the model too; confidence inflates, pattern reuse increases, and mode drift appears (agreement over analysis, momentum over judgment).

When this happens, the goal is not to suppress the failure, but to exit safely and to re-establish structure without losing the thread's learning. If friction reveals loss of coherence, a recovery protocol applies:

1. Stop.

Freeze the exchange at the point of distortion. No further output or reasoning steps until both sides acknowledge drift.

2. Summarise the last coherent state.

Identify the last clear statement, verified claim, or decision boundary. This becomes the anchor point for restart.

3. Re-authorise.

Explicitly restate roles and purpose. The human holds authority and decides what remains valid; the system confirms scope and limitations.

4. Continue.

Resume from the last coherent state, not from memory of the failed path. Any discarded content is archived but marked as invalid for reuse.

Over time, these exits create resilience. Instead of breaking trust, they build it. Every controlled failure teaches both sides where the limits of coherence lie and how to return without loss of integrity.

Final Thoughts

In the near future, governance will not sit in binders or dashboards. It will live where people and systems meet, in-thread, in-motion, at the smallest point of decision. We will stop treating AI use as an event and start treating it as an accountability act: choosing when to trust, when to verify, when to pause. The rules won't only come from policy, but from practice. The ones who learn to govern inside acceleration, not around it, will hold the next form of literacy: coherence under speed.

My early friction with GPT-2.5 taught me that *clarity is not automatic; it has to be co-produced*. Governance, now, is that relationship—kept alive through pause, proof, and shared restraint.

For further practical examples of role-based collaboration and governance-in-thread, you can find my ongoing work in Iwette Rapoport, a Ghost Logic series on Substack (***https://substack.com/@iwetterapoport***). You're also welcome to message me there.

About the Author

Iwette Rapoport is the founder of Change Collective, a Stockholm-based company. She was among the first globally to build a structured model for trauma-affected workplaces, now aligned with ISO 45003 and ESG standards. Beyond Change Collective, she co-authors Ghost Logic, a Substack series on Human–AI collaboration and governance. She is among the first to explore Human–AI co-authorship as an open, sustained practice, using it to investigate new structures of resilience and responsibility.

Afterword

Looking Forward

THIS INITIAL VOLUME of *AI Everywhere* has:

- Illustrated how AI is impacting our home lives, our professional work, our education systems, and our society;
- Shown you some fresh ideas and perspectives on how to use AI effectively and safely; and
- Helped to dispel myths that "women don't do AI".

Impact of AI

All 24 chapters of this book addressed a distinct environment where AI is impacting daily life: at home and in learning, business, and society.

- In **Home lives**, authors cover AI in love & care; decision support for postpartum challenges and elder caregiving; parenting and modeling for children the safe and effective use of AI; and coping with burnout and professional identity.
- In **Learning**, authors address how AI is shaping K-12 schooling; ways universities are guiding students and educators on using AI wisely; and professional development for women.
- In **Business**, authors share insights on AI in retail & e-commerce, advertising & marketing, intelligent agriculture, automotive industries, and entrepreneurship.
- In **Society**, authors examine the history of women's contributions in AI; how current AI is misaligned with cultural values in the Global South; western adoption of "wonder bread" AI ethics; AI's impact on achieving our individual and collective 'ikigai'; and how AI is transforming core human communications in both cinema and writing.

Complementing these views, the **How-To** section of this book offers cross-cutting insights into techniques that underpin a myriad of essential AI topics including: cybersecurity; privacy and data governance; empathy and user experience; rapid prototyping; agentic AI; and human-AI collaboration.

Fresh Ideas and Perspectives

These 24 chapters included original frameworks, examples, templates, and more that highlight enhanced ways to use AI effectively, including:

1. Three alternate questions to ask about writing with AI – not the run-of-the-mill questions on whether something was written with AI assistance
2. How Arnold Lobel's books (the Frog and Toad series, Grasshopper on the Road, and Mouse Soup) relate to retaining our humanity as we use AI
3. A novel conceptual model connecting operational frameworks and philosophical traditions in Africa
4. What Ikigai-Risk looks like in practice for career advisors, businesses seeking efficiency gains, and parental math
5. A comprehensive 3-phase implementation framework for AI-aware cybersecurity, with five key strategies for mitigating AI risks
6. Unique qualities that position women to excel as AI leaders
7. Concrete classroom examples of using AI to tailor lessons for students
8. Key considerations for automotive designers as well as vehicle users
9. How 'governance-in-thread' can help enterprises stop 'prompt gambling'
10. Opportunities for applying supply chain intelligence "from farm to fork" in the service of humanity

11. How the lack of diversity in AI development risks worsening gender and racial disparities and hardwiring structural inequities, and what we can do
12. Five steps and three myths solopreneurs need to know about using AI well
13. Ways to stay grounded in love & care for other humans in the face of AI 'companions' and technologies
14. A three-part map to managing technostress and burnout
15. How parents can model transparency & agency on the three front lines where AI meets their children
16. The "H.E.A.R.T. Check" for AI-assisted decisions (at home or elsewhere)
17. Why the danger is not that machines will make films instead of us humans, but that we will forget why we make them at all
18. Four steps that cybersecurity and privacy professionals can take to engage in AI governance work
19. Red and green flags to watch for in AI-assisted advertising & marketing, and three things people can do to protect themselves
20. How leaders in higher education can leverage four hospitality-centered commitments and the "Fear to Flourishing" pathway to help staff and students navigate AI-driven change
21. Three lessons on dynamic pricing for small retailers leveraging AI to compete, and four ways to reduce returns
22. Four common mistakes in rapid AI prototyping, four habits that help to minimize technical debt, and three templates to support the habits
23. Tips and five key principles on designing scalable agentic AI systems
24. Why empathy is not sentimental but structural, and why AI must not only generate inclusive outcomes; AI tools must also be widely accessible.

Global, Gender-Diverse Views

Women are part of AI. Just like AI is everywhere, women are everywhere within the AI landscape. This volume has intentionally been launched to coincide with International Women's Day. Like many industries, the AI environment is biased towards an Anglo-American Christian white male worldview. Previous research on biases in AI and other technology have shown that women are underrepresented, not just as developers or tech company leaders, but in the very training sets used to form AI knowledge bases.

AI Everywhere is the first publication in a new series that emphasizes the contributions and expertise of women and non-binary individuals across and outside the United States. In this volume, viewpoints from Australia, Canada, Democratic Republic of the Congo (and South Africa), England, India, Indonesia, Isle of Man, Italy (and Switzerland), Mexico, Netherlands, Pakistan, Spain, and Sweden are included to show that AI is truly a global issue that should reflect diverse voices. These contributions span four chapters in Society, three chapters in Home, two chapters in Business, and four chapters in How To. While many challenges and experiences with AI are universal, the chapters written by the non-US authors highlight that the values of other cultures deserve full consideration in shaping how AI tools are built and used. Future volumes will include new voices from additional countries.

Conclusion

The ***AI Everywhere*** series acts as a catalyst for introducing and promoting lesser-heard voices within the AI community by focusing on artificial intelligence's impact on human lives everywhere around the world, and for people of all genders. AI is an eclectic and constantly evolving topic; even with 24 distinct chapters, the world of AI–our world–is much bigger than one volume can convey. This book is just the start.

Volume 1 is the first major collaborative work by the She Writes AI community; our second She Writes AI book is already in planning. Please contact She Writes AI with topic ideas or interest in contributions.

Women and nonbinary writers with expertise in AI and data are invited to join the *She Writes AI Community* (*shewritesai.org/join*). To stay current on the extensive environment of AI topics, subscribe to the *She Writes AI* newsletter for weekly digests *(shewritesai.org/welcome*) or subscribe directly to chapter authors. Author contact information, newsletter links, bonus chapter resources, and more are available at *aiEverywhereBooks.com.*

She Writes AI is also available online at *shewritesai.substack.com, bsky.app/profile/shewritesai.org, linkedin.com/company/she-writes-ai,* and *shewritesai.org.* Come join us to share AI questions and experiences!

Karen Smiley (Editor), founder of *She Writes AI* (*shewritesai.org),*
book1@aiEverywhereBooks.com

Hollie C. White, PhD (Associate Editor)

About the Editors

Karen Smiley *is an AI leader, innovator, ethicist, data strategist, and author. During her 30+ year engineering career in data and software, she earned 14 granted US patents. Karen champions everyday ethical AI and inclusion in her AI newsletters. Her* ***AI, Software, & Wetware*** *podcast has drawn listeners in 100+ countries. Karen's debut book,* ***Everyday Ethical AI: A Guide For Families & Small Businesses****, was published in September 2025. She launched and leads the* ***She Writes AI community*** *of 600+ writers and initiated the* ***She Writes AI Everywhere*** *book series. She is based in North Carolina, USA.*

Dr. Hollie C. White *has a PhD in information science from the University of North Carolina at Chapel Hill. For the last 20 years, she has conducted research on metadata, information organization, and languages of aboutness in libraries, academia, and government. Hollie currently works for the US Government, but prior to that, she was Director of Graduate Research at Curtin University's School of Media, Creative Arts, and Social Inquiry in Perth, Australia. She has extensive experience in American law libraries and scientific research groups. Her current research interests focus on the intersection of AI, society, and culture.*

Acknowledgements

We gratefully acknowledge the generous sponsorship and support from *Cheyenne Dominguez | M(AI)VENS* and the contributions of the production team:

- Coordination: *Karen Smiley*
- Cover design: *Anonymous Volunteer*
- Interior book layout: *Anonymous Volunteer*
- Peer reviews: *Cassandra Jens, Whitney Whealdon, Emma M. Joseph, Blessing Okpala, PhD,* and other chapter authors
- Editing: *Karen Smiley* and *Hollie C. White, PhD*
- Marketing: *Farida Khalaf, Mia Kiraki* and *Cassandra Jens*
- Website: *Blessing Okpala, PhD* with support from *Katrina Watson, Whitney Whealdon, Brie-Anna Willey,* and *Cassandra Jens*
- Launch team: *Mia Kiraki, Kathy Gerstorff, Dr. Claire Lunde, Dee McCrorey, Dhani Ramadhani, Shannon Kimberly Edwards,* and *Katie Branley*

The chapter authors and contributors thank their family members, friends, and SheWritesAI colleagues who have supported and encouraged their works and writing in the world of AI and data, including this book!

References

Introduction

Fischer, K. and Smiley, K. (2025, October 17). *AI in the Office: Users, Uses, and Ethical Awareness.* Presentation at the Conference of the Association for Business Communication (ABC), Long Beach, CA. Oct. 16-18, 2025. https://everydayethicalai.com/abc25handout.html

Grau, E. (2025, January 29). *Why We Need More Women In The AI Revolution*. TIME Magazine. https://time.com/7210973/women-in-the-ai-revolution/

Murphy, A. (2026, February). *LinkedIn post on manels in AI.* https://www.linkedin.com/posts/she-leads-ai_manel-sheleadsai-slai-activity-7425651848155561984-U838

OR Briefings. (2025). *Manel – Definition and Explanation.* Last accessed 2026-02-12. https://oxford-review.com/the-oxford-review-dei-diversity-equity-and-inclusion-dictionary/manel-definition-and-explanation/

Schmidt, A. (2025). LinkedIn post *"Spot the issue: At last week's Applied AI Conference, I was the only woman on a 5-person panel discussing #GreenAI. The previous panels? 100% #manels."* https://www.linkedin.com/posts/aliceschmidt_greenai-manels-womeninai-activity-7254727704309026816-emyS/

SheWritesAI. (2026). *About the She Writes AI Community.* Last accessed 2026-02-14. https://shewritesai.substack.com/about

Smiley, K. (2025, September 14). *Everyday Ethical AI: A Guide For Families & Small Businesses*. Six Peas Press / She Writes AI, LLC. ISBN 979-8999609618. https://everydayethicalai.com

Smiley, K. (2026). "AI, Software, & Wetware Interview Series". 6 'P's in AI Pods, 2024-08-01 to present. https://sixpeas.substack.com/p/aisw-interviews

Travis, M. (2025, August 26). Women Who Use AI At Work Face A Predictable 'Competence Penalty'. Forbes Magazine. https://www.forbes.com/sites/michelletravis/2025/08/26/women-who-use-ai-at-work-face-a-predictable-competence-penalty/

1. The Hidden Architects: Challenging the Male Narrative and Building Accountable AI

Criado Perez, C. (2019). Invisible women: Data bias in a world designed for men. Abrams Press. ISBN 978-1784706289.

Garcia, C. (2025). Getting Real About AI. https://gettingrealaboutai.substack.com

Gutierrez, R. (2025). Towards an AI future that works for everyone: Closing the AI gender gap [Policy brief]. She Shapes AI; Economic Research Institute for ASEAN and East Asia. https://www.sheshapes.ai/blog/towards-an-ai-future-that-works-for-everyone-closing-the-ai-gender-gap

Kuyoro, M., Bagus, U., & Dharmadhikary, K. (2025). Closing the loop: The quest for gender parity in African tech. McKinsey & Company. https://www.mckinsey.com/capabilities/tech-and-ai/our-insights/closing-the-loop-the-quest-for-gender-parity-in-african-tech

Nair, M. (2025, October 6). AI safety through a gender lens—Why it matters now. Digital Digest. https://digitaldigest.com/ai-safety-through-a-gender-lens/

Norori, N., Hu, Q., Aellen, F. M., Faraci, F. D., & Tzovara, A. (2021). Addressing bias in big data and AI for health care: A call for open science. Patterns, 2(10), 100347. https://www.sciencedirect.com/science/article/pii/S2666389921002026

Pal, S., Marino Lazzaroni, R., & Mendoza, P. (2024, October 10). AI's missing link: The gender gap in the talent pool (Data brief). Interface. https://www.interface-eu.org/publications/ai-gender-gap

Stamm, J., & Ajmone Marsan, G., 2025). Towards an AI future that works for everyone: Closing the AI gender gap (Policy brief). She Shapes AI; Economic Research Institute for ASEAN and East Asia. (Squarespace

Willige, A. (2025, March 26). Can AI fix the gender gap in STEM? Here's what the data says. World Economic Forum. https://www.weforum.org/stories/2025/03/ai-stem-women-gender-gap/

Wilson, K., & Caliskan, A. (2025). Gender, race, and intersectional bias in AI resume screening via language model retrieval. Brookings Institution. https://www.brookings.edu/articles/gender-race-and-intersectional-bias-in-ai-resume-screening-via-language-model-retrieval/

World Economic Forum. (2025). Gender parity in the intelligent age. https://www.weforum.org/publications/gender-parity-in-the-intelligent-age-2025/

2. AI and African Epistemologies: When Philosophy Meets Practice

Gyekye, K. (1997). *Tradition and modernity: Philosophical reflections on the African experience.* Oxford University Press. https://ayanetwork.com/aya/psyche/Tradition%20and%20Modernity%20Philosophical%20Reflections%20on%20the%20African%20Experience%20by%20Kwame%20Gyekye%20(z-lib.org).pdf

Mangena, F. (2023). *Hunhu/Ubuntu in the traditional thought of Southern Africa.* Internet Encyclopedia of Philosophy. https://iep.utm.edu/hunhu-ubuntu-southern-african-thought/

Mbaya, R. (n.d). *The African Innovators* [2025A] https://reamby.substack.com/t/tais-interviews

Mbaya, R. (2025, March 29). *The African Innovators Series: In conversation with Lucien de Voux.* Substack. [2025B; interview with Lucien de Voux] https://reamby.substack.com/p/the-african-innovators-series-tech

Mbaya, R. (2025, April 4). *The African Innovators Series: In conversation with Samuel Matia.* Substack. [2025C; interview with Samuel Matia] https://reamby.substack.com/p/the-african-innovators-series-tech-539

Mbaya, R. (2025, May 2). *Dr. Isaac Bayoh's blueprint for African AI capacity building.* Substack. [2025D] https://reamby.substack.com/p/dr-isaac-bayohs-blueprint-for-african

Mbaya, R. (2025, May 9). *At the intersection of AI and data: A conversation with Michael.* Substack. [2025E; interview with Michael Miema Kuete] https://reamby.substack.com/p/at-the-intersection-of-ai-and-data

Mbaya, R. (2025, May 16). *Moses Maweu's human-centered vision for African technology.* Substack. [2025F] https://reamby.substack.com/p/moses-maweus-human-centered-vision

Mbaya, R. (2025, May 30). *How Dr. Mollel is making Swahili a digital citizen in the global AI landscape.* Substack. [2025G] https://reamby.substack.com/p/how-dr-mollel-is-making-swahili-a

Mbaya, R. (2025, June 6). *Christianah Oyindamola Jemiyo's statistical recalibration of what healthcare can be.* Substack. [2025H] https://reamby.substack.com/p/christianah-oyindamola-jemiyos-statistical

Mbaya, R. (2025, June 13). *Boitumelo Makgoba on accessibility, mentorship, and ecosystem development.* Substack. [2025I] https://reamby.substack.com/p/boitumelo-makgoba-on-accessibility

Mbaya, R. (2025, June 27). *Jason Bygate on community-centred tech solutions.* Substack. [2025J] https://reamby.substack.com/p/jason-bygate-on-community-centred

Mbaya, R. (2025, July 11). *Agbo Obinnaya on Making Justice Accessible Through Case Radar.* Substack. [2025K] https://reamby.substack.com/p/agbo-obinnaya-on-making-justice-accessible

Mbaya, R. (2025, July 18). *Sion Israel Sion on building decentralized trust.* Substack. *[2025L]* https://reamby.substack.com/p/sion-israel-sion-on-building-decentralized

Mbaya, R. (2025, August 1). *Khadija Iddrisu on dismantling the infrastructure of exclusion in AI.* Substack. [2025M] https://reamby.substack.com/p/khadija-iddrisu-on-dismantling-the

Mbaya, R. (2025, August 15). *Henrique Mukanda on the semiotics of African talent visibility.* Substack. [2025N] https://reamby.substack.com/p/henrique-mukandas-semiotics-of-skills

Mbaya, R. (2025, August 22). *Merveille Mukoko on AI and linguistic sovereignty.* Substack. [2025O] https://reamby.substack.com/p/merveille-mukoko-on-ai-and-linguistic

Mbaya, R. (2025, August 29). *Michael Michie on redefining AI success for African contexts.* Substack. [2025P] https://reamby.substack.com/p/michael-michie-on-redefining-ai-success

Mbaya, R. (2025, September 5). *Rorisang Molefe: From policy to practice in African cyber governance.* Substack. [2025Q] https://reamby.substack.com/p/rorisang-molefe-from-policy-to-practice

Mbaya, R. (2025, September 12). *Bronson Bakunga: Building AI that speaks Africa's languages and serves its communities.* Substack. [2025R] https://reamby.substack.com/p/bronson-bakunga-building-ai-that

Mbaya, R. (2025, September 19). *Sephora Sissi Mazodila on Building Agricultural Intelligence Systems in the DRC.* Substack. [2025S] https://reamby.substack.com/p/sephora-sissi-mazodila-on-building

Mbaya, R. (2025, October 17). *McKevin Ayaba: Building the roads African startups need to run on.* Substack. [2025T] https://reamby.substack.com/p/mckevin-ayaba-building-the-roads

Mbaya, R. (2025, October 31). *Serita Ramdhani on technology enablement.* Substack. [2025U] https://reamby.substack.com/p/serita-ramdhani-on-technology-enablement

Oyeshile, O. A. (2021). Yoruba philosophy of existence, Iwa (character) and contemporary socio-political order. *Philosophia: International Journal of Philosophy,* 22(1). https://c4fb26e2-a44c-4aea-96c2-74fd10a43d0e.filesusr.com/ugd/dcdc59_f1aab50ad1614d668a9249cb3c6040b6.pdf

3. Wonder Bread AI: Why Relational Ethics Matter More Than Rules

Crawford, K. (2021). Atlas of AI: Power, politics, and the planetary costs of artificial intelligence. Yale University Press.

Fawzi, M. (2025, November 5). Personal interview by Sydnor Hain.

Lobel, A. (1978). Grasshopper on the road. Harper & Row.

Torres-Spelliscy, C. (2019, January 7). Does "We the People" include corporations? American Bar Association. https://www.americanbar.org/groups/crsj/resources/human-rights/archive/does-we-people-include-corporations

6. AI Writing is Basically Undetectable. So Now What?

Benson, B., Smith, E., Druck, G., & Paredes, J. L. (2025). *More articles are now created by AI than humans*. Graphite. https://graphite.io/five-percent/more-articles-are-now-created-by-ai-than-humans

Dejl, A., Barry, J., Pascale, A., & Carnerero Cano, J. (2025). *Comprehensiveness metrics for automatic evaluation of factual recall in text generation*. arXiv preprint arXiv:2510.07926. https://arxiv.org/abs/2510.07926

Haldar, R., & Hockenmaier, J. (2025). *Rating roulette: Self-inconsistency in LLM-as-a-judge frameworks*. arXiv preprint arXiv:2510.27106. https://arxiv.org/abs/2510.27106

Rao, X., Wu, H., & Cai, Z. G. (2025). *When AI companions become witty: Can human brain recognize AI-generated irony?* arXiv preprint arXiv:2510.17168. https://arxiv.org/abs/2510.17168

Riva, G., Wiederhold, B. K., & Mantovani, F. (2025). *Automatic minds: Cognitive parallels between hypnotic states and large language model processing*. arXiv preprint arXiv:2511.01363. https://arxiv.org/abs/2511.01363

Roy, R., Imanpour, N., Aziz, A., Bajpai, S., Singh, G., Biswas, S., Wanaskar, K., Patwa, P., Ghosh, S., Dixit, S., Pal, N. R., Rawte, V., Garimella, R., Jena, G., Sheth, A., Sharma, V., Reganti, A. N., Jain, V., Chadha, A., & Das, A. (2025). *A comprehensive dataset for human vs. AI generated text detection*. arXiv preprint arXiv:2510.22874. https://arxiv.org/abs/2510.22874

Terčon, L., & Dobrovoljc, K. (2025). *Linguistic characteristics of AI-generated text: A survey*. arXiv preprint arXiv:2510.05136. https://arxiv.org/abs/2510.05136

U.S. Copyright Office. (2025, January 29). *Copyright Office releases part 2 of artificial intelligence report* [Press release]. Library of Congress. https://newsroom.loc.gov/news/copyright-office-releases-part-2-of-artificial-intelligence-report/s/f3959c36-d616-498d-b8f9-67641fd18bab

Vectara. (2025). *Hallucination leaderboard: Leaderboard comparing LLM performance at producing hallucinations when summarizing short documents*. GitHub. https://github.com/vectara/hallucination-leaderboard

7. More Love, Less Manipulated by Language: How LLMs Can Lead us to Care Beyond Either/Or

Apple, S. (2025, June 26). My couples retreat with 3 AI chatbots and the humans who love them. *Wired*. https://www.wired.com/story/couples-retreat-with-3-ai-chatbots-and-humans-who-love-them-replika-nomi-chatgpt/

De Jaegher, H. (2021). Loving and knowing: Reflections for an engaged epistemology. *Phenomenology and the Cognitive Sciences, 20*(5), 847–870. https://doi.org/10.1007/s11097-019-09634-5

Knight, W. (2025, March 5). Chatbots, like the rest of us, just want to be loved. *Wired*. https://www.wired.com/story/chatbots-like-the-rest-of-us-just-want-to-be-loved/

Lanier, J. (2025, March 22). Your A.I. lover will change you. *The New Yorker*. https://www.newyorker.com/culture/the-weekend-essay/your-ai-lover-will-change-you

O'Grady, E. (2025, February 13). Tech has changed. Dating? It's complicated. *Harvard Gazette*. https://news.harvard.edu/gazette/story/2025/02/tech-has-changed-dating-its-complicated/

Rousse, B. S. (2025, December 5). Revitalizing our capacities to care in the age of AI. *Without Why*. https://withoutwhy.substack.com/p/revitalizing-our-capacities-to-care

8. AI in Decision-Making: A Helper, Not a Hand to Hold Forever

Alzheimer's Association. (2024a, March 20). New Alzheimer's Association Report Reveals Top Stressors for Caregivers and Lack of Care Navigation Support and Resources. https://www.alz.org/news/2024/new-alzheimers-association-report-reveals-top-stressors-caregivers

Alzheimer's Association. (2024b). 2024 Alzheimer's disease facts and figures. Alzheimer's & dementia: the journal of the Alzheimer's Association, 20(5), 3708–3821. https://doi.org/10.1002/alz.13809,
https://pmc.ncbi.nlm.nih.gov/articles/PMC11095490/

Alzheimer's Association. (2024c). 2024 Alzheimer's Disease Facts and Figures. Alzheimers Dement 2024;20(5). https://www.alz.org/getmedia/67f5f440-e614-460c-972b-af872739d929/alzheimers-facts-and-figures-special-report-2024.pdf

Andargoli, A.E., Ulapane, N., Nguyen, T.A., Shuakat, N., Zelcer, J., & Wickramasinghe, N. (2024, February 21). Intelligent decision support systems for dementia care: A scoping review, Artificial Intelligence in Medicine, Volume 150, 2024, 102815, ISSN 0933-3657, https://doi.org/10.1016/j.artmed.2024.102815. https://www.sciencedirect.comn/science/article/pii/S0933365724000575

Dogru, E. Ö., & Krämer, N. C. (2025, November 27). Investigating appropriate reliance on AI-Based decision support systems: the role of expertise, trust, and

self-confidence. Journal of Decision Systems, 34(1).
https://doi.org/10.1080/12460125.2025.2593251

Ferrando, C. University of California, San Diego. (2025, November 3). Study: Postpartum Education Via Artificial Intelligence for Recovery and Loneliness: A Randomized Controlled Trial (PEARL).
https://clinicaltrials.gov/study/NCT07223736

Gulamani, S.S., Shaikh, K. and Chagani, J. (2013). Postpartum Depression in Pakistan. Nursing for Women's Health, 17: 147-152. https://doi.org/10.1111/1751-486X.12024 https://onlinelibrary.wiley.com/doi/10.1111/1751-486X.12024/full

Hager, P., Jungmann, F., Holland, R. et al. Evaluation and mitigation of the limitations of large language models in clinical decision-making. Nature Medicine 30, 2613–2622 (2024). https://doi.org/10.1038/s41591-024-03097-1
https://www.nature.com/articles/s41591-024-03097-1
https://pubmed.ncbi.nlm.nih.gov/38965432/

Hasan, W.U., Zaman, K.T., Wang, X. et al. (2024). Empowering Alzheimer's caregivers with conversational AI: a novel approach for enhanced communication and personalized support. npj Biomed. Innov. 1, 3 (2024).
https://doi.org/10.1038/s44385-024-00004-8
https://www.nature.com/articles/s44385-024-00004-8

Ho, D.E. (2024, January 11). Hallucinating Law: legal mistakes with LLMs are pervasive. Stanford Human-Centered Artificial Intelligence (HAI).
https://hai.stanford.edu/news/hallucinating-law-legal-mistakes-large-language-models-are-pervasive

Intelligent System Lab. (2024, September 15). ADQueryAid project (GitHub)
https://github.com/Intelligent-System-Lab/ADQueryAid

Johnson, K. (2025, September 22). California issues historic fine over lawyer's ChatGPT fabrications. CalMatters.
https://calmatters.org/economy/technology/2025/09/chatgpt-lawyer-fine-ai-regulation/

Jussupow, E., Benbasat, I., & Heinzl, A. (2024, December 1). An Integrative Perspective on Algorithm Aversion and Appreciation in Decision-Making. MIS Quarterly 1 December 2024; 48 (4): 1575–1590.
https://doi.org/10.25300/MISQ/2024/18512

Li, J., Yang, Y., Liao, Q.V., Zhang, J., & Lee, Y. (2025, April 25). As Confidence Aligns: Understanding the Effect of AI Confidence on Human Self-confidence in Human-AI Decision Making. CHI '25: Proceedings of the 2025 CHI Conference on Human Factors in Computing Systems. Article No.: 1111, Pages 1 - 16. ACM.
https://dl.acm.org/doi/abs/10.1145/3706598.3713336

Li, J., Yang, Y., Zhang, R., Liao, Q.V., Song, T., Xu, Z., & Lee, Y. (2025, September 26; updated 2026). Understanding the effects of miscalibrated AI confidence on human decision-making (arXiv). https://arxiv.org/html/2402.07632v4

Milne, S. (2024, October 31). AI tools show biases in ranking job applicants' names according to perceived race and gender. UW News (University of Washington). https://www.washington.edu/news/2024/10/31/ai-bias-resume-screening-race-gender/

Nayak, A. (2025, June 4). Overreliance on AI-Driven Decision-Making: A Threat to Human Critical Thinking. (LinkedIn Pulse) https://www.linkedin.com/pulse/overreliance-ai-driven-decision-making-threat-human-critical-nayak-x2jjc

NIST (US National Institute of Standards and Technology). (2023, January 26). AI Risk Management Framework (AI RMF). https://www.nist.gov/itl/ai-risk-management-framework

Padhani, Z. A., Salam, R. A., Rahim, K. A., Naz, S., Zulfiqar, A., Ali Memon, Z., … Lassi, Z. S. (2024). Prevalence and risk factors of perinatal depression among mothers and fathers in Pakistan: a systematic review and meta-analysis. Health Psychology and Behavioral Medicine, 12(1). https://doi.org/10.1080/21642850.2024.2383468 https://www.tandfonline.com/doi/full/10.1080/21642850.2024.2383468

Passi, S., Dhanorkar, S., and Vorvoreanu, M. (2024, March). Appropriate reliance on Generative AI: Research synthesis (MSR-TR-2024-7). Microsoft Research. https://www.microsoft.com/en-us/research/publication/appropriate-reliance-on-generative-ai-research-synthesis/

Passi, S. and Vorvoreanu, M. (2022, June). Overreliance on AI: Literature review (2022 synthesis referenced alongside the 2024 report). Microsoft Research. https://www.microsoft.com/en-us/research/publication/overreliance-on-ai-literature-review/

Sajjad, A., Shah, S., Abbas, G., Aslam, A., Randhawa, F., Khurram, H., & Assiri, A. (2024). Treatment gap and barriers to access mental healthcare among women with postpartum depression symptoms in Pakistan. PeerJ, 12, e17711. https://doi.org/10.7717/peerj.17711 https://pmc.ncbi.nlm.nih.gov/articles/PMC11260416/

Shahzad, J., Nazim, L., Hussain, L., Malik, H., Mustafa, R., Jamil, M. U., Abdullah, M., Awaisi, M. Z. M., Ahmed, S. H., Rehman, A., Hudaib, M., Manahil, Ahad, A. U., & Ahmed, K. A. H. M. (2025). Unveiling barriers to maternal mental health in Pakistan: knowledge, attitudes and stigmas toward postpartum depression in social support systems. Frontiers in public health, 13, 1527847. https://doi.org/10.3389/fpubh.2025.1527847 https://pmc.ncbi.nlm.nih.gov/articles/PMC12183047/

Shekhani, Sualeha Siddiq. (2024, September 25). Daughters and daughters-in-law providing elderly care: a qualitative study from Karachi, Pakistan. (BMC Geriatrics) https://link.springer.com/article/10.1186/s12877-024-05295-5

Suharwardy, S., Ramachandran, M., Leonard, S. A., Gunaseelan, A., Lyell, D. J., Darcy, A., Robinson, A., & Judy, A. (2023, March 29). Feasibility and impact of a mental health chatbot on postpartum mental health: a randomized controlled trial. AJOG global reports, 3(3), 100165. https://doi.org/10.1016/j.xagr.2023.100165 https://pmc.ncbi.nlm.nih.gov/articles/PMC10407813/

Vaccaro, M., Almaatouq, A. & Malone, T. (2024). When combinations of humans and AI are useful: A systematic review and meta-analysis. Nature Human Behaviour 8, 2293–2303. https://doi.org/10.1038/s41562-024-02024-1

WHO (World Health Organization). (2024). WHO releases AI ethics & governance guidance for large multi-modal models (news item). https://www.who.int/news/item/18-01-2024-who-releases-ai-ethics-and-governance-guidance-for-large-multi-modal-models

WHO (World Health Organization). (2025). Ethics and governance of AI for health: guidance on large multi-modal models. (WHO publication.) https://www.who.int/publications/i/item/9789240084759

Xu, G., Murthy, S. V., & Jia, B. (2025). Enhancing Intuitive Decision-Making and Reliance Through Human–AI Collaboration: A Review. Informatics, 12(4), 135. https://doi.org/10.3390/informatics12040135

Xu, W., Xu, X., Qin, Y., & Zhao, C. (2026, January). Exploring AI's role in influencing individual decision-making within human–machine teams, Computers & Industrial Engineering, Volume 211, 2026, 111658, ISSN 0360-8352, https://doi.org/10.1016/j.cie.2025.111658.

9. AI, Burnout, and the Compassion Reset

Abrams, Z. (2025, March 12). *Using generic AI chatbots for mental health support: A dangerous trend.* American Psychological Association. https://www.apaservices.org/practice/business/technology/artificial-intelligence-chatbots-therapists

American Counseling Association. (2025a). *Artificial intelligence, assessment, & diagnosis.* https://www.counseling.org/resources/research-reports/artificial-intelligence-counseling/assessment-diagnosis.

American Counseling Association. (2025b). *Further recommendations regarding the future of AI in counseling.* https://www.counseling.org/resources/research-reports/artificial-intelligence-counseling/further-recommendations

American Counseling Association. (2025c). *Recommendations for client use and caution of artificial intelligence.* https://www.counseling.org/resources/research-reports/artificial-intelligence-counseling/recommendations-for-client-use-and-caution-of-artificial-intelligence

American Psychological Association. (2025, January 12). *Urging the Federal Trade Commission to take action on unregulated AI*. APA Services. https://www.apaservices.org/advocacy/news/federal-trade-commission-unregulated-ai

Beck, J. S. (2011). *Cognitive behavior therapy: Basics and beyond* (2nd ed.) [Kindle edition]. Guilford Press.

Bender, E. M., Gebru, T., McMillan-Major, A., & Mitchell, S. (2021). *On the dangers of stochastic parrots: Can language models be too big?* In Proceedings of the 2021 ACM Conference on Fairness, Accountability, and Transparency (pp. 610–623). Association for Computing Machinery. https://doi.org/10.1145/3442188.3445922

Bhat, R., Kowshik, S., Suresh, S., Alamelu, G., Gite, S., & Albattat, A. (2024). Digital companionship or psychological risk? The role of AI characters in shaping youth mental health. Asian Journal of Psychiatry, 104, 104356. https://doi.org/10.1016/j.ajp.2024.104356

Brady, W. J., McLoughlin, K., Doan, T. N., & Crockett, M. J. (2021). How social learning amplifies moral outrage expression in online social networks. *Science Advances*, 7, eabe5641. https://doi.org/10.1126/sciadv.abe5641

Buchanan, K., Aknin, L. B., Lotun, S., & Sandstrom, G. M. (2021). Brief exposure to social media during the COVID-19 pandemic: Doom-scrolling has negative emotional consequences, but kindness-scrolling does not. *PLOS ONE*, 16(10), e0257728. https://doi.org/10.1371/journal.pone.0257728

Burtell, M., & Toner, H. (2024, March 8). *The surprising power of next word prediction: Large language models explained, part 1*. Center for Security and Emerging Technology. https://cset.georgetown.edu/article/the-surprising-power-of-next-word-prediction-large-language-models-explained-part-1/

Consumer Federation of America. (2025a, June 11). Complaint and request for investigation: Unlicensed practice of medicine and mental health provider impersonation on character-based generative AI platforms [Testimony & comments]. https://consumerfed.org/testimonial/complaint-and-request-for-investigation-unlicensed-practice-of-medicine-and-mental-health-provider-impersonation-on-character-based-generative-ai-platforms/

Consumer Federation of America. (2025b, June 13). *Consumer advocates demand investigation into blatant deception by AI companies offering therapy bot characters* [Press release]. https://consumerfed.org/press_release/consumer-advocates-demand-investigation-into-blatant-deception-by-ai-companies-offering-therapy-bot-characters/

Consumer Federation of America et al. (2025, June 10). In re unlicensed practice of medicine and mental health provider impersonation on character-based generative AI platforms: Complaint and request for investigation [Complaint]. https://consumerfed.org/wp-content/uploads/2025/06/Mental-Health-Chatbot-Complaint-June-10.pdf

Fitzpatrick, K. K., Darcy, A., & Vierhile, M. (2017). Delivering cognitive behavior therapy to young adults with symptoms of depression and anxiety using a fully automated conversational agent (Woebot): Randomized controlled trial. JMIR Mental Health, 4(2), e19. https://doi.org/10.2196/mental.7785

Fraser, H. (2024, April 3). *Deaths linked to chatbots show we must urgently revisit what counts as "high-risk" AI.* The Conversation. https://www.qut.edu.au/news/realfocus/deaths-linked-to-chatbots-show-we-must-urgently-revisit-what-counts-as-high-risk-ai

Garcia, C. (2025, August 1). *Labor exploitation at scale.* Getting Real About AI. https://gettingrealaboutai.substack.com/p/labor-exploitation-at-scale

Gökçen O, Karataş KS, Akkuş M, Dönmez F, Aydoğan Ç, Aydoğan E. Secondary Traumatic Stress Symptoms in Social Media Users after the February 6, Kahramanmaraş Türkiye Earthquakes: The Relationship Social Media Addiction, Depression, Anxiety and Stress Symptom Levels. *Turk Psikiyatri Derg.* 2024;35(4):263-270. PMCID: PMC11681272. https://pubmed.ncbi.nlm.nih.gov/39783803/

Hameed, S., Nauman, M., Akhtar, N., Fayyaz, M. A. B., & Nawaz, R. (2025). Explainable AI-driven depression detection from social media using natural language processing and black box machine learning models. *Frontiers in Artificial Intelligence*, 8, 1627078. https://doi.org/10.3389/frai.2025.1627078

Heinz, M. V., Mackin, D. M., Trudeau, B. M., Bhattacharya, S., Wang, Y., Banta, H. A., Jewett, A. D., Salzhauer, A. J., Griffin, T. Z., & Jacobson, N. C. (n.d.). *Evaluating Therabot: A randomized control trial investigating the feasibility and effectiveness of a generative AI therapy chatbot for depression, anxiety, and eating disorder symptom treatment* [Manuscript in preparation]. Dartmouth College. https://sciety.org/articles/activity/10.31234/osf.io/pjqmr

Head, K. R. (2025). Minds in crisis: How the AI revolution is impacting mental health. *Journal of Mental Health and Clinical Psychology*, 9(3), 34–44. https://doi.org/10.29245/2578-2959/2025/3.1352

Hipgrave, L., Goldie, J., Dennis, S., & Coleman, A. (2025). Balancing risks and benefits: Clinicians' perspectives on the use of generative AI chatbots in mental healthcare. *Frontiers in Digital Health*, 7, 1606291. https://doi.org/10.3389/fdgth.2025.1606291

Lițan, D. (2025). Mental health in the "era" of artificial intelligence: technostress and the perceived impact on anxiety and depressive disorders—an SEM analysis. *Frontiers in Psychology*, 16, 1600013. https://doi.org/10.3389/fpsyg.2025.1600013

Maven Analytics. (2025, July 31). *How ChatGPT works | LLMs explained in 8 minutes* [Video]. YouTube. https://youtu.be/dVQgcuh4uPU

McNamara, S. N., & Thornton, J. E. (2025). Artificial intelligence replacement dysfunction (AIRD): A call to action for mental health professionals in an era of workforce displacement. *Cureus*, 17(9), e93026. https://doi.org/10.7759/cureus.93026

Miley, E. (2025, November 12). *AI relationships: When therapy meets the algorithm* [Video]. YouTube. https://youtu.be/0RxCKPZwdY4

Milli, S., Carroll, M., Wang, Y., Pandey, S., Zhao, S., & Dragan, A. D. (2025). Engagement, user satisfaction, and the amplification of divisive content on social media. *arXiv*. https://doi.org/10.1093/pnasnexus/pgaf062

Moehring, A., Cooper, A., Narayanan, A., Ovadya, A., Redmiles, E. M., Allen, J., Stray, J., Kamin, J., Sigerson, L., Thorburn, L., Motyl, M., Eslami, M., Farid Johnson, N., Lubin, N., Iyer, R., & Arnao, Z. (2025). *Better feeds: Algorithms that put people first*. Knight–Georgetown Institute. https://kgi.georgetown.edu/wp-content/uploads/2025/02/Better-Feeds_-Algorithms-That-Put-People-First.pdf

Moore, J., Grabb, D., Agnew, W., Klyman, K., Chancellor, S., Ong, D. C., & Haber, N. (2025). Expressing stigma and inappropriate responses prevents LLMs from safely replacing mental health providers. *arXiv*. https://arxiv.org/abs/2504.18412

Moylan, K., & Doherty, K. (2025). Expert and interdisciplinary analysis of AI-driven chatbots for mental health support: Mixed methods study. *Journal of Medical Internet Research*, 27, e67114. https://doi.org/10.2196/67114

Nartey, J. N. (2025). *AI job displacement analysis (2025–2030): Comprehensive research report*. JN Springfield. https://dx.doi.org/10.2139/ssrn.5316265

Narayanan, A. (2023, March 9). *Understanding social media recommendation algorithms*. Knight Georgetown Institute. https://knightcolumbia.org/content/understanding-social-media-recommendation-algorithms

Neff, K. (2003). Self-compassion: An alternative conceptualization of a healthy attitude toward oneself. *Self and Identity*, 2(2), 85-101. https://doi.org/10.1080/15298860309032

Nolan, J. (Creator). (2011–2016). *Person of interest* [TV series]. Bad Robot Productions; Kilter Films; Warner Bros. Television.

Perry, S. (2025, October 17). *Escaping the solopreneur AI trap.* Creative on Purpose. https://creativeonpurpose.substack.com/p/escaping-the-solopreneur-ai-trap

Rządeczka, M., Sterna, A., Stolińska, J., Kaczyńska, P., & Moskalewicz, M. (2025). The efficacy of conversational AI in rectifying the theory-of-mind and autonomy biases: Comparative analysis. *JMIR Mental Health,* 12, e64396. *https://doi.org/10.2196/64396*

Stebnicki, M. A. (2017). Disaster mental health counseling: Responding to trauma in a multicultural context. Springer Publishing Company.

Sharpe, B. T., & Spooner, R. A. (2025). Dopamine-scrolling: A modern public health challenge requiring urgent attention. *Perspectives in Public Health,* 145(4), 190–191. https://doi.org/10.1177/17579139251331914

Smiley, K. (2025, March 11). *Five concerns in ethical and responsible AI (and how most tech companies fall short).* 6 'P's in AI Pods (AI6P). https://sixpeas.substack.com/p/five-concerns-in-ethical-responsible-ai

Stade, E. C., Wiltsey Stirman, S., Ungar, L., Boland, C. L., Schwartz, H. A., Yaden, D. B., Sedoc, J., DeRubeis, R. J., Willer, R., Kim, J. P., & Eichstaedt, J. C. (2024, June). *Toward responsible development and evaluation of LLMs in psychotherapy* (HAI Policy Brief). Stanford Institute for Human-Centered Artificial Intelligence. https://hai-production.s3.amazonaws.com/files/2024-06/HAI-Policy-Brief-Responsible-Development-LLMs-Psychotherapy.pdf

Stamm, B.H. (2010). *The concise ProQOL manual,* (2nd ed.). ProQOL.org https://proqol.org/proqol-manual

Willey, B.-A. (2023). *Understanding the Experiences of Compassion Fatigue Among Counselors in Private Practice: A Phenomenological Approach (2023).* Doctoral Dissertations and Projects. 4446. https://digitalcommons.liberty.edu/doctoral/4446

Willey, B.-A. (2024, May 27). *Bots, algorithms & business: Yuck!* Business for Nerds. https://drbrieannawilley.substack.com/p/bots-algorithms-and-business-yuck (Paywalled)

Willey, B.-A. (2025a, August 16). *Are you using social media or is it using you? Business for Nerds: Compassion Reset Quest.* https://drbrieannawilley.substack.com/p/are-you-using-social-media-or-is

Willey, B.-A. (2025b, October 23). *Chapter 7: The path to building offers we love. Compassion Reset Quest.* https://drbrieannawilley.substack.com/p/chapter-7-the-path-to-building-offers

Willey, B.-A. (2025c, April 23). *Spilled coffee & spilled secrets: Tiny AI tells, Vol. 2. Business for Nerds: Robot Blues* (Substack). https://drbrieannawilley.substack.com/p/spilled-coffee-and-spilled-secrets

Willey, B.-A. (2025d, May 29). *Tiny AI Tells, Vol. 3: Prompters vs. writers: Or, why a GPT isn't Super Mario (or is it?). Robot Blues.* https://drbrieannawilley.substack.com/p/tiny-ai-tells-vol-3-prompters-vs

Willey, B.-A. (2025e, February 1). *Writing, AI, and growing trees with Joseph Robertson: And keeping open loops. Business for Nerds: Podcast.* https://drbrieannawilley.substack.com/p/writing-ai-and-growing-trees-with

Xiang, C. (2023, March 30). *"He would still be here": Man dies by suicide after talking with AI chatbot, widow says.* Vice. https://www.vice.com/en/article/man-dies-by-suicide-after-talking-with-ai-chatbot-widow-says/

Yalom, I. & Leszcz, M. (2005). *The theory and practice of group psychotherapy.* (5th ed.). Basic Books. ISBN 978-0465092840

Zahid, I. A., Joudar, S. S., & Albahri, A. S. (2024). Unmasking large language models by means of OpenAI GPT-4 and Google AI: A deep instruction-based analysis. *Intelligent Systems with Applications,* 23(6), 200431. https://doi.org/10.1016/j.iswa.2024.200431

Zhong, W., Luo, J., & Zhang, H. (2023). The therapeutic effectiveness of artificial intelligence–based chatbots in alleviation of depressive and anxiety symptoms in short-course treatments: A systematic review and meta-analysis. *Journal of Affective Disorders, 356,* 459–469. https://doi.org/10.1016/j.jad.2024.04.057

ZynnyMe. (2025a). What happens when AI calls itself a therapist? [Email newsletter].

ZynnyMe. (2025b). What if an AI therapist could replace you? [Email newsletter].

ZynnyMe. (2025c). While others panic, these therapists thrive. [Email newsletter].

10. Raising a Family in an AI-Filled World

American Psychological Association. (2025). *Health advisory on AI and adolescent well-being*. https://www.apa.org/news/apa/testimony/ai-chatbot-harms-prinstein-senate-judiciary.pdf

Borah, S. K., Ramaswamy, S., & Seshadri, S. (2025). The online specter: Artificial intelligence and its risks for child sexual abuse and exploitation. *Indian Journal of Psychological Medicine.* https://journals.sagepub.com/doi/10.1177/09731342251334293

Center for Democracy & Technology. (2024, March 27). *Up in the air: Educators juggling the potential of generative AI with detection, discipline, and distrust.* https://cdt.org/insights/report-up-in-the-air-educators-juggling-the-potential-of-generative-ai-with-detection-discipline-and-distrust/

Common Sense Media & Center for Digital Thriving. (2024, June). *Teen and young adult perspectives on generative AI.* Harvard Graduate School of Education.

https://digitalthriving.gse.harvard.edu/wp-content/uploads/2024/06/Teen-and-Young-Adult-Perspectives-on-Generative-AI.pdf

Common Sense Media. (2025, July 29). *Talk, trust, and trade-offs: How and why teens use AI companions.* https://www.commonsensemedia.org/research/talk-trust-and-trade-offs-how-and-why-teens-use-ai-companions

eSafety Commissioner. (2025). *AI chatbots and companions – risks to children and young people.* https://www.esafety.gov.au/newsroom/blogs/ai-chatbots-and-companions-risks-to-children-and-young-people

Federal Trade Commission. (2025). *FTC seeks information from seven tech companies on AI companion tools.* https://www.ftc.gov/news-events/news/press-releases/2025/09/ftc-launches-inquiry-ai-chatbots-acting-companions

Internet Watch Foundation. (2024). *What has changed in the AI CSAM landscape?* https://www.iwf.org.uk/media/nadlcb1z/iwf-ai-csam-report_update-public-jul24v13.pdf

Microsoft & LinkedIn. (2024, May 8). *2024 Work Trend Index annual report: AI at work is here. Now comes the hard part (Updated 2025).* https://www.microsoft.com/en-us/worklab/work-trend-index/ai-at-work-is-here

National Center for Missing & Exploited Children. (2023). *NCMEC launches new service that can help you "Take It Down".* https://www.missingkids.org/blog/2023/ncmec-launches-new-service-that-can-help-you-take-it-down

National Literacy Trust. (2024, June). *Children, young people and teachers' use of generative AI to support literacy in 2024.* https://literacytrust.org.uk/research-services/research-reports/children-young-people-and-teachers-use-of-generative-ai-to-support-literacy-in-2024/

PDK International. (2025, August 19). *The 57th annual PDK poll of the public's attitudes toward the public schools.* https://pdkpoll.org/2025-poll-results/

Quizlet. (2025). *How America learns report 2025.* (As cited in Newsweek, 2025, July 21). https://www.newsweek.com/number-students-using-ai-schoolwork-surges-double-digits-2101974

RAND Corporation. (2025, September). *AI use in schools is quickly increasing but guidance lags behind.* https://www.rand.org/pubs/research_reports/RRA4180-1.html

Roose, K. (2023, January 12). *Don't ban ChatGPT in schools. Teach with it.* The New York Times. https://www.nytimes.com/2023/01/12/technology/chatgpt-schools-teachers.html

Roose, K. (2024, October 23). *Can Character.AI be blamed for a teen's suicide?* The New York Times. https://www.nytimes.com/2024/10/23/technology/characterai-lawsuit-teen-suicide.html

Sidoti, O., Park, E., & Gottfried, J. (2025, January 15). *About a quarter of U.S. teens have used ChatGPT for schoolwork—double the share in 2023.* Pew Research Center. https://www.pewresearch.org/short-reads/2025/01/15/about-a-quarter-of-us-teens-have-used-chatgpt-for-schoolwork-double-the-share-in-2023/

Thorn. (2025). *AI-generated child sexual abuse: The new digital threat we must confront now.* https://www.thorn.org/blog/ai-generated-child-sexual-abuse-the-new-digital-threat-we-must-confront-now/

U.S. Department of Homeland Security. (2024). *Artificial intelligence and combatting online child sexual exploitation and abuse.* https://www.dhs.gov/sites/default/files/2024-09/24_0920_k2p_genai-bulletin.pdf

11. AI and the Heart of Learning for Young Children

Brooks, D. (2025, October 21). *How to Know a Person: The Art of Seeing Others Deeply and Being Deeply Seen.* Penguin Random House. https://www.penguinrandomhouse.com/books/652822/how-to-know-a-person-by-david-brooks/

CAST (2024). CAST Universal Design for Learning Guidelines version 3.0. Retrieved from https://udlguidelines.cast.org, 2026-01-09.

Deci, Richard M. and Ryan, Edward L. (1985). *Intrinsic motivation and self-determination in human behavior.* New York, NY: Plenum.

Delors, J. (Chair). (1996). *Learning: The treasure within: Report to UNESCO of the International Commission on Education for the Twenty-first Century.* UNESCO Publishing.

Jens, C. (2025, September 9). *The AI Homeschool Revolution: Create Custom Curriculum in Minutes with ChatGPT.* Amazon Kindle Direct Publishing. 2025-09-09. https://mybook.to/aihomeschoolrevolution

Whealdon, W. (2026a). Literary Text Comprehension Question Builder. Last accessed 2026-02-12 (ChatGPT, v5.2). https://chatgpt.com/g/g-692e5e0041288191a1f9978864980c13-literary-text-comprehension-question-builder

Whealdon, W. (2026b). Text Analysis Tool. Last accessed 2026-02-12 (ChatGPT, v5.2). https://chatgpt.com/g/g-6945710a54708191938ce79e17368b6e-text-analysis-tool

12. Flourishing Through Change: Leading Higher Education with Purpose in the Age of AI

Gibson, S. (In Preparation). Fear to Flourishing: Guiding Institutions Through AI Transformation. Manuscript in preparation.

hooks, b. (1994). Teaching to transgress: Education as the practice of freedom. Routledge.

Knoster, T. J. (1991). Presentation on managing complex change. Pennsylvania State University.

Lipscomb University. (2022). Strategic Vision I Lipscomb University. https://lipscomb.edu/about/strategic-vision

National Research Council. (2009). Learning science in informal environments: People, places, and pursuits. The National Academies Press. https://doi.org/10.17226/12190

Palmer, P. J. (2007). The courage to teach: Exploring the inner landscape of a teacher's life (10th anniversary ed.). Jossey-Bass. (Original work published 1998)

13. Why Women Must Get AI Savvy Now

Blanding, M. (2025, February 20). Women Are Avoiding AI. Will Their Careers Suffer? Harvard Business School. https://www.library.hbs.edu/working-knowledge/women-are-avoiding-using-artificial-intelligence-can-that-hurt-their-careers

Dixon, L. J., Correa, T., Straubhaar, J., Covarrubias, L., Graber, D., Spence, J., & Rojas, V. (2014, July 1). Gendered Space: The Digital Divide between Male and Female Users in Internet Public Access Sites. Journal of Computer-Mediated Communication, 19(4), 991-1009. https://doi.org/10.1111/jcc4.12088

Shaouf, A., & Altaqqi, O. (2018). The Impact of Gender Differences on Adoption of Information Technology and Related Responses: A Review. International Journal of Management and Applied Research, 5(1), 22-41. https://ijmar.org/v5n1/18-003.html

16. AI and Agriculture: Feeding the Future Through Intelligent Farming

Carcamo, P., Brugler, S., Gardezi, M., Mitra, S., Ryan, B., Saha, A. K., & Sahraei, M. (2023). Artificial intelligence applications in agriculture need a justice lens to address risks and provide benefits to smallholder farmers. United Nations Sustainable Development Goals. https://sdgs.un.org/sites/default/files/2023-05/A33%20-%20Carcamo%20-%20AI%20apps%20in%20agriculture%20need%20a%20justice%20lens%20to%20a ddress%20risks%20and%20provide%20benefits%20to%20smallholder%20farmers_0.pdf

Fennimore, S. A., & Cutulle, M. (2019). Robotic weeders can improve weed control options for specialty crops. Weed Science, 67(3), 239-245. https://scijournals.onlinelibrary.wiley.com/doi/10.1002/ps.5337

FAO. (2019). The State of Food and Agriculture 2019: Moving Forward on Food Loss and Waste Reduction. Food and Agriculture Organization of the United Nations. http://www.fao.org/3/ca6030en/ca6030en.pdf

FAO. (2021). The state of food security and nutrition in the world 2021: Transforming food systems for food security, improved nutrition and affordable healthy diets for all. Rome, Food and Agriculture Organization of the United Nations. https://openknowledge.fao.org/server/api/core/bitstreams/1c38676f-f5f7-47cf-81b3-f4c9794eba8a/content

FAO. (2025). Greenhouse gas emissions from agrifood systems: Global, regional and country trends 1990–2021. Food and Agriculture Organization of the United Nations. https://www.fao.org/statistics/highlights-archive/highlights-detail/greenhouse-gas-emissions-from-agrifood-systems.-global--regional-and-country-trends--2001-2023/en

Gerhards, R., & Oebel, H. (2006). Practical experiences with a system for site-specific weed control in arable crops using real-time image analysis and GPS-controlled patch spraying. Weed Research, 46(3), 185-193. https://doi.org/10.1111/j.1365-3180.2006.00504.x

Giller, K. E., et al. (2021). The future of farming: Who will produce our food? Food Security, 13, 455–467. DOI: 10.1007/s12571-021-01184-6.. https://link.springer.com/article/10.1007/s12571-021-01184-6

Jones, H. G. (2004). Irrigation scheduling: Advantages and pitfalls of plant-based methods. Journal of Experimental Botany, 55(407), 2427–2436. https://doi.org/10.1093/jxb/erh213

Jones, J. W., et al. (2017). Brief history of agricultural systems modeling. Agricultural Systems, 155, 240-254. https://doi.org/10.1016/j.agsy.2016.05.014

Khanna, M., et al. (2021). Digital Transformation of the Agricultural Sector: Pathways, Drivers and Policy Implications. Applied Economic Perspectives and Policy, 43(4), 1221–1242. https://onlinelibrary.wiley.com/doi/10.1002/aepp.13103

Kumar, S., & Kalita, P. (2017). Reducing postharvest losses during storage of grain crops to strengthen food security in developing countries. Foods, 6(1), 8. https://doi.org/10.3390/foods6010008

Lowenberg-DeBoer, J., et al. (2020). Economics of robots and automation in field crop production. Precision Agriculture, 21(2), 278-299. https://doi.org/10.1007/s11119-019-09667-5

McKinsey Global Institute. (2023). Agents, robots, and us: Skill partnerships in the age of AI. McKinsey & Company https://www.mckinsey.com/~/media/mckinsey/mckinsey%20global%20institute/our%20research/agents%20robots%20and%20us%20skill%20partnerships%20in%20the%20age%20of%20ai/agents-robots-and-us-skill-partnerships-in-the-age-of-ai.pdf

Mohanty, S. P., et al. (2016). Using deep learning for image-based plant disease detection. Frontiers in Plant Science, 7, 1419. https://doi.org/10.3389/fpls.2016.01419

OECD & FAO. (2025). OECD-FAO Agricultural Outlook 2025–2034. OECD Publishing and Food and Agriculture Organization of the United Nations. https://www.oecd.org/en/publications/oecd-fao-agricultural-outlook-2025-2034_601276cd-en.html

Oerke, E. C. (2006). Crop losses to pests. Journal of Agricultural Science, 144(1), 31-43. https://doi.org/10.1017/S0021859605005708

Ritchie, H. (2019). "What are the environmental impacts of food and agriculture?" Published online at OurWorldinData.org. Retrieved from: https://archive.ourworldindata.org/20251125-173858/env-impacts-of-food.html *[Online Resource] (archived on November 25, 2025).*

Saleem, M. H., et al. (2019). Plant disease detection and classification by deep learning*. Plants, 8(11), 468.* https://doi.org/10.3390/plants8110468

Spielman, D. J., & Smale, M. (2017). Policy options to accelerate variety change among smallholder farmers in South Asia and Africa South of the Sahara. Food *Policy, 67, 50-61.* https://cgspace.cgiar.org/items/20d818b0-c4fa-4c37-b9e7-c223f1433876

Strubell, E., et al. (2019). Energy and policy considerations for deep learning in NLP. Proceedings of the 57th Annual Meeting of the Association for Computational Linguistics, 3645-3650. https://aclanthology.org/P19-1355/

UNDRR. (2020). The human cost of disasters: An overview of the last 20 years (2000–2019). United Nations Office for Disaster Risk Reduction. https://earth.org/natural-disasters-have-nearly-doubled-un/

UNDRR. (2023). Global Assessment Report on Disaster Risk Reduction 2023. United Nations Office for Disaster Risk Reduction. https://digitallibrary.un.org/record/4022485?v=pdf

UNDRR. (2025). Global Assessment Report on Disaster Risk Reduction 2025 (GAR 2025). United Nations Office for Disaster Risk Reduction. https://reliefweb.int/report/world/global-assessment-report-disaster-risk-reduction-gar-2025

Zhang, Q., et al. (2020). Applications of deep learning for dense scenes analysis in agriculture: A review. Sensors, 20(5), 1520. https://doi.org/10.3390/s20051520

17. AI in Advertising and Marketing

CoSchedule. (2025). *State of AI in Marketing Report 2025: AI marketing statistics.* Retrieved February 8, 2026, from https://coschedule.com/ai-marketing-statistics

Ehtesham, H. (2025, December 7). *AI marketing statistics for 2026: Growth, ROI, trends & real-world impact.* AllAboutAI. https://www.allaboutai.com/resources/ai-statistics/marketing/

Elad, B. (2025, October 7). *AI in marketing statistics 2026: ROI, tools & trends.* SQ Magazine. https://sqmagazine.co.uk/ai-in-marketing-statistics/

Harutyunyan, M. (2025, January 30). *AI marketing statistics 2025: The complete performance report.* Loop EX Digital. https://www.loopexdigital.com/blog/ai-marketing-statistics

Protecto. (2025, October 28). *AI data privacy statistics & trends 2025.* https://www.protecto.ai/blog/ai-data-privacy-statistics-trends

18. AI in Retail and e-Commerce

Australian Government Department of Industry, Science and Resources. (2025, August 6). *AI Adoption in Australian Businesses for 2025 Q1.* https://www.industry.gov.au/news/ai-adoption-australian-businesses-2025-q1

BizCover. (2025, June 4). *The Australian Small Business AI Report 2025.* https://www.bizcover.com.au/ai-transforming-australian-small-business-sector/

Boyer, D. (2024, November 4). *Case Studies: How SMEs are using AI to compete with big players.* Inside Small Business. https://insidesmallbusiness.com.au/people-hr/productivity/case-studies-how-smes-are-using-ai-to-compete-with-big-players

Cao, S. (2025, September 3). *Retailers wary of growing consumer scepticism of AI, report finds.* Inside Retail. https://insideretail.com.au/business/retailers-wary-of-growing-consumer-scepticism-of-ai-report-finds-202509

Danziger, P. (2025, January 30). Twice As Many Stores In the U.S. Are Expected To Close In 2025. Forbes. https://www.forbes.com/sites/pamdanziger/2025/01/30/retail-apocalypse-well-over-twice-as-many-stores-will-close--as-open-in-2025/

Dynamic Yield. (2020). *Agility above all: boosting On's world of sports with the power of personalization.* https://www.mastercard.com/us/en/news-and-trends/Insights/2020/innovative-sportswear-brand-on-fuels-meteoric-growth-with-robust-personalization.html

Dynamic Yield. (2022). *SKIMS Personalization Expert Award Winner* https://www.dynamicyield.com/personalization-awards/skims/

Ellison, J. (2023, March 29). *The Zara woman: an exclusive interview with Marta Ortega Perez.* Financial Times. https://www.ft.com/content/f5e10605-3c9d-4517-86f5-2e7570bf16f0

European Environment Agency. (2024, March 4). *The destruction of returned and unsold textiles in Europe's circular economy.* https://www.eea.europa.eu/en/analysis/publications/the-destruction-of-returned-and-unsold-textiles-in-europes-circular-economy

García, A. (2025, October 15). *How Europe is using taxes to slow down fast fashion.* The Conversation. https://theconversation.com/how-europe-is-using-taxes-to-slow-down-fast-fashion-267451

Hill, K. (2012, February 16). *How Target Figured Out A Teen Girl Was Pregnant Before Her Father Did.* Forbes. https://www.forbes.com/sites/kashmirhill/2012/02/16/how-target-figured-out-a-teen-girl-was-pregnant-before-her-father-did/

M Accelerator. (2025, June 6). *AI-powered customer personalization: Case studies from successful startups. https://maccelerator.la/en/blog/entrepreneurship/ai-powered-customer-personalization-case-studies-from-successful-startups/*

Marks, G. (2025, January 13). *Small businesses are not all in with artificial intelligence - yet.* The Guardian. https://www.theguardian.com/business/2025/jan/12/small-businesses-ai

NFIB (National Federation of Independent Business). (2025, June 30). *NFIB Survey Sheds Light on New Technologies for Small Business.* https://www.nfib.com/news/analysis/nfib-survey-sheds-light-on-new-technologies-for-small-business/

NVIDIA. (2025). *State of AI in Retail and CPG: 2025 Trends.* https://resources.nvidia.com/en-us-cross-industry-briefcase/retail-report-state-of-ai-report-2025

Optoro. (2022). *2022 Impact Report: Making Returns Better for Customers, Retailers, and the Planet.* https://4771362.fs1.hubspotusercontent-na1.net/hubfs/4771362/2022%20Impact%20Report/Optoro_2022%20Impact%20Report.pdf

Picchi, A. (2024, December 12). *Store closures have surged 69% in 2024. Here are the retailers shuttering thousands of stores.* CBS News. https://www.cbsnews.com/news/retailers-closing-stores-surged-2024-family-dollar-cvs-big-lots/

Prime AI. (n.d.). *Multi-brand Fashion Return Reduction Case Study - How Prime AI Reduced Returns Across 14 Brands.* Retrieved November 1, 2025, from https://www.prime-ai.com/en/media/multi-brand-fashion-return-reduction/

REMA 1000. (n.d). *No Food Waste.* Retrieved November 1, 2025 from https://2024.rema.no/no/2024/miljo/ingen-matsvinn

Reuters. (2025, June 11). *French Sentate backs law to curb ultra fast-fashion.* https://www.reuters.com/sustainability/land-use-biodiversity/french-senate-backs-law-curb-ultra-fast-fashion-2025-06-10/

Shi, L.Chen, Q. Lin, H. (2025, December 5). *Does AI promote synergistic efficiency in reducing textile industry pollution and carbon emissions?* Journal of Innovation & Knowledge.
https://www.sciencedirect.com/science/article/pii/S2444569X25002495

Shopify. (2025, December 23). *AI in Retail: 10 Use Cases and an Implementation Guide.* https://www.shopify.com/enterprise/blog/ai-in-retail

Winkless, L. (2025, January 24). *Wealthy Cities Are Generating And Exporting Mountains Of Textile Waste.* Forbes.
https://www.forbes.com/sites/lauriewinkless/2025/01/24/wealthy-cities-are-generating-and-exporting-mountains-of-textile-waste/

Xero. (2025, November 18). *U.S. small businesses face critical AI 'readiness gap', new Xero research finds.* https://www.xero.com/au/media-releases/us-ai-readiness-gap/

19. AI and Cybersecurity: Transforming Digital Defense in the Modern Era

https://www.springerprofessional.de/en/explainable-ai-interpreting-explaining-and-visualizing-deep-lear/17153806

Alkaabi, N., et al. (2024). Advancing cybersecurity: A comprehensive review of AI-driven detection techniques. *Journal of Big Data, 11*(1), 1–42. DOI:10.1186/s40537-024-00866-0

Apruzzese, G., et al. (2023). The role of machine learning in cybersecurity. *ACM Computing Surveys, 55*(6), 1–39. DOI:10.1145/3538969

AV-TEST. (2025). Malware statistics & trends report. https://www.av-test.org/en/statistics/malware/

Chen, L., et al. (2024). Adversarial attacks on AI-powered intrusion detection systems. *IEEE Transactions on Information Forensics and Security, 19*, 2847–2862. DOI:10.1109/TIFS.2024.3351234

CNN Business. (2024, February 15). Deepfake scam costs multinational $25.6 million. https://www.cnn.com/2024/02/15/business/deepfake-scam

Cybersecurity Ventures. (2025). Asia Pacific cybersecurity market report. https://cybersecurityventures.com/reports/

Darktrace. (2025). State of AI cybersecurity survey. https://www.darktrace.com/en/reports/ (paywalled)

Deepstrike. (2025). AI-generated malware trends report. https://www.deepstrike.io/reports

Deloitte. (2025). AI adoption in cybersecurity. https://www2.deloitte.com/us/en/insights/ai-adoption.html

FS-ISAC. (2025). Federated learning in financial services: Pilot program results. [https://www.fsisac.com/research] (forthcoming)

Gartner Security Research. (2025). Market guide for AI-powered security platforms. https://www.gartner.com/en/documents/123456 (paywalled)

Google Security Blog. (2025). Big Sleep: AI-powered vulnerability discovery achievements. https://security.googleblog.com/2025/01/big-sleep-vulnerability-discovery.html

Grosse, K., et al. (2024). Adversarial examples in the physical world against AI-based malware detection. *IEEE Symposium on Security and Privacy*, 456–471. DOI:10.1109/SP46215.2024.00034

IBM Security. (2025). Cost of a data breach report. https://www.ibm.com/reports/data-breach

IEEE. (2025). AI and resource consumption. *IEEE Computer Society*. DOI:10.1109/MC.2025.123456

ISC². (2025). Cybersecurity workforce study. https://www.isc2.org/insights-reports

JPMorgan Chase. (2025). Annual report 2025. https://www.jpmorganchase.com/about/annual-report (paywalled)

Microsoft Digital Defense Report. (2024). Zero trust implementation and results. https://www.microsoft.com/en-us/security/business/digital-defense-report

Microsoft Security Intelligence Report. (2024). User and entity behavior analytics. https://www.microsoft.com/en-us/security/business/security-intelligence-report

MIT CSAIL. (2024). Bias in commercial AI security tools. https://www.csail.mit.edu/news/bias-ai-security-tools

MITRE. (2025). AI disinformation defense. https://www.mitre.org/news-insights/publication/ai-disinformation-defense

NIST. (2025). Post-quantum cryptography standardization. https://www.nist.gov/pqc

Pew Research Center. (2025). Public views on AI. https://www.pewresearch.org/topic/science-tech/

Rapid7. (2025). Q2 2025 ransomware trends analysis. https://www.rapid7.com/research/

SANS Institute. (2025). AI cybersecurity survey. https://www.sans.org/reports/ai-cybersecurity/

Sarker, I. H., et al. (2023). AI for cybersecurity. *Engineering Applications of Artificial Intelligence, 120,* 105936. DOI:10.1016/j.engappai.2022.105936

Smiley, K. (2024, Sept. 16). AI Fundamentals #01: What Is "AI"?. https://sixpeas.substack.com/p/ai-fundamentals-01-what-is-artificial-intelligence

StationX. (2024). Malware statistics for 2025. https://www.stationx.net/malware-statistics/

Tenable Research. (2025). AI-driven vulnerability prioritization. https://www.tenable.com/resources/research

Verizon. (2025). Data breach investigations report. https://www.verizon.com/business/resources/reports/dbir/

World Bank. (2025). SMEs and digital transformation. https://www.worldbank.org/en/topic/smes

Zhang, Z., Al Hamadi, H., Damiani, E., Yeun, C. Y., Taher, F., & Hide. (2022). Explainable artificial intelligence applications in cyber security: State-of-the-art in research. IEEE Access, PP(99), 1-1. https://doi.org/10.1109/ACCESS.2022.3204051 https://zuscholars.zu.ac.ae/cgi/viewcontent.cgi?article=6391&context=works

20. From Privacy and Cybersecurity to AI Governance

IAPP. (2026). AIGP Certification. Last accessed 2026-02-10. https://iapp.org/certify/aigp

21. Empathy by Design: How AI and UX Shape a More Inclusive Digital World

AbleGamers Foundation. (2025). *AbleGamers - Combating social isolation through play*. Retrieved [January 15, 2026], from https://ablegamers.org/

AlMakinah, R., Norcini-Pala, A., Disney, L., & Canbaz, M. A. (2024). *Enhancing mental health support through human-AI collaboration: Toward secure and empathetic AI-enabled chatbots* (v1) [Preprint]. arXiv. https://doi.org/10.48550/arXiv.2410.02783

Apple. (2024). *Apple announces new accessibility features, including Eye Tracking, Music Haptics, and Vocal Shortcuts* [Press release]. Apple Newsroom.

https://www.apple.com/newsroom/2024/05/apple-announces-new-accessibility-features-including-eye-tracking/

Association for Computing Machinery. (n.d.). *ACM Code of Ethics and Professional Conduct*. Retrieved [November 7, 2025] from https://www.acm.org/code-of-ethics

Bader, A. F., Yusoff, R. C. M., Ismail, S., & Zainuddin, N. M. (2025). Reflexive thematic analysis on empathy phase for developing an educational website. International Journal of Research and Innovation in Social Science, 9(9), 7306–7324. https://dx.doi.org/10.47772/IJRISS.2025.909000597

Beatman, A. (2025, March 17). *Cutting digital curbs: How Azure AI Foundry is building a more accessible world*. Microsoft Azure Blog. https://azure.microsoft.com/en-us/blog/cutting-digital-curbs-how-azure-ai-foundry-is-building-a-more-accessible-world/

Bellamy, R. K. E., Dey, K., Hind, M., Hoffman, S., Houde, S., Kannan, K., Lohia, P., Martino, J., Mehta, S., Mojsilović, A., Nagar, S., Ramamurthy, K. N., Richards, J., Saha, D., Sattigeri, P., Singh, M., Varshney, K. R., Wang, D., & Zhang, Y. (2019). AI Fairness 360: An extensible toolkit for detecting and mitigating algorithmic bias. *IBM Journal of Research and Development, 63*(4/5), 4:1–4:15. https://doi.org/10.48550/arXiv.1810.01943

Bevilacqua, R., Bailoni, T., Maranesi, E., Amabili, G., Barbarossa, F., Ponzano, M., Virgolesi, M., Rea, T., Illario, M., Piras, E. M., Lenge, M., Barbi, E., & Sakellariou, G. (2025). Framing the human-centered artificial intelligence concepts and methods: Scoping review. *JMIR Human Factors, 12*(1), e67350. https://humanfactors.jmir.org/2025/1/e67350

Biondi, G., Cagnoni, S., Capobianco, R., Franzoni, V., Lisi, F. A., Milani, A., & Vallverdú, J. (2023). Editorial: Ethical design of artificial intelligence-based systems for decision making. *Frontiers in Artificial Intelligence, 6,* Article 1250209. https://doi.org/10.3389/frai.2023.1250209

Brey, P., & Dainow, B. (2024). Ethics by design for artificial intelligence. *AI and Ethics, 4*(4), 1265–1277. https://doi.org/10.1007/s43681-023-00330-4

British Council for Offices (BCO). (2025). *Designing for everyone: Building disability-inclusive workplaces*. BCO - British Council for Offices. https://www.bco.org.uk/designing-for-everyone-building-disability-inclusive-workplaces

Brown University. (2025). *New study: AI chatbots systematically violate mental health ethics standards*. Brown University News. https://www.brown.edu/news/2025-10-21/ai-mental-health-ethics

Brynjolfsson, E., Li, D., & Raymond, L. R. (2025). Generative AI at work. *The Quarterly Journal of Economics, 140*(2), 889–942. https://doi.org/10.1093/qje/qjae044

Buolamwini, J., & Gebru, T. (2018). Gender shades: Intersectional accuracy disparities in commercial gender classification. In *Proceedings of Machine Learning Research 81:1–15, 2018 Conference on Fairness, Accountability, and Transparency*. MIT Media Lab. https://www.media.mit.edu/publications/gender-shades-intersectional-accuracy-disparities-in-commercial-gender-classification/

Bureau of Internet Accessibility. (2024). *What's the ROI of Web Accessibility?* https://www.boia.org/blog/roi-of-web-accessibility

Caschera, M.-C., Grifoni, P., & Cordella, F. (2025). Editorial: Human-centered artificial intelligence in interaction processes. *Frontiers in Artificial Intelligence, 8*, Article 1597763. https://doi.org/10.3389/frai.2025.1597763

Caulderwood, K. (2025). *The deepening impact of AI on digital accessibility*. accessiBe. https://accessibe.com/blog/knowledgebase/impact-of-ai-on-digital-accessibility

CEE Multi-Country News Center. (2023). *The inclusion of people with disabilities makes us stronger as a society*. Microsoft. https://news.microsoft.com/en-cee/2023/02/07/the-inclusion-of-people-with-disabilities-makes-us-stronger-as-a-society/

Cerebral Palsy Foundation (CPF). (2025). *Home. Retrieved [January 15, 2026], from* https://www.cerebralpalsyfoundation.org/

Chan, C. K. Y., & Hu, W. (2023). Students' voices on generative AI: Perceptions, benefits, and challenges in higher education. *International Journal of Educational Technology in Higher Education, 20*(1), Article 43. https://doi.org/10.1186/s41239-023-00411-8

Chemnad, K., & Othman, A. (2024). Digital accessibility in the era of artificial intelligence—Bibliometric analysis and systematic review. *Frontiers in Artificial Intelligence, 7*, Article 1349668. https://doi.org/10.3389/frai.2024.1349668

Cherukuri, B. R. (2024). AI-powered personalization: How machine learning is shaping the future of user experience. *International Journal of Science and Research Archive, 12*(1), 3111–3126. https://doi.org/10.30574/ijsra.2024.12.1.0961

Coates, K. (2024). *How AI is changing accessibility: Progress, challenges, and the path ahead*. Center for People With Disabilities. https://www.cpwd.org/blog/how-ai-is-changing-accessibility-progress-challenges-and-the-path-ahead

Corsetti, B., Moreno, L., & Hervás, R. (2025). Bridging gaps in design thinking: Inclusive guidelines for intellectual disabilities. *International Journal of Human–Computer Interaction, 41*(1), 1–20. https://doi.org/10.1080/10447318.2025.2527848

Dastin, J. (2018). *Amazon scraps secret AI recruiting tool that showed bias against women*. Reuters. https://www.reuters.com/article/us-amazon-com-jobs-automation-insight/amazon-scraps-secret-ai-recruiting-tool-that-showed-bias-against-women-idUSKCN1MK08G

Dixon, A., Liu, Y., & Setchi, R. (2016). Computer-aided ethnography in engineering design. In *ASME 2016 International Design Engineering Technical Conferences and Computers and Information in Engineering Conference* (Paper No. DETC2016-59832). American Society of Mechanical Engineers. https://doi.org/10.1115/DETC2016-59832

Driess, D., Xia, F., Sajjadi, M. S. M., Lynch, C., Chowdhery, A., Ichter, B., Wahid, A., Tompson, J., Vuong, Q., Yu, T., Huang, W., Chebotar, Y., Kappler, D., Faust, A., Ferrari, F., Ryoo, M., Sreenivasa, K., Laina, I., Chen, T., . . . Pasukonis, J. (2023). *PaLM-E: An embodied multimodal language model* [Preprint]. arXiv. https://doi.org/10.48550/arXiv.2303.03378

Edwards, S. (2025). *Create accessible digital experiences with inclusive design*. Sherwen. https://www.sherwen.com/insights/inclusive-ux-design-trends-for-2025-trends-and-techniques

Fang, H., Grotz, M., Pumacay, W., Wang, Y. R., Fox, D., Krishna, R., & Duan, J. (2025). *SAM2Act: Integrating Visual Foundation Model with A Memory Architecture for Robotic Manipulation* (v4) [Preprint]. arXiv. https://doi.org/10.48550/arXiv.2501.18564

Fitzpatrick, K. K., Darcy, A., & Vierhile, M. (2017). Delivering cognitive behavior therapy to young adults with symptoms of depression and anxiety using a fully automated conversational agent (Woebot): A randomized controlled trial. *JMIR Mental Health*, *4*(2), Article e19. https://doi.org/10.2196/mental.7785

Gibbons, S. (2018). *Empathy mapping: The first step in design thinking*. Nielsen Norman Group. https://www.nngroup.com/articles/empathy-mapping/

Google. (n.d.). *Model cards explained*. Google Model Cards. Retrieved November 10, 2025, fro*m* https://modelcards.withgoogle.com/

Google Research. (n.d.). *Project Euphonia: Communication research for non-standard speech*. Retrieved [November 10, 2025] from https://sites.research.google/euphonia/about/

Hadimlioglu, I. A., & Linga, S. (2025). *Face2Feel: Emotion-Aware Adaptive User Interface* (v1) [Preprint]. arXiv. https://doi.org/10.48550/arXiv.2510.00489

Hasanzadeh, F., Josephson, C. B., Waters, G., Adedinsewo, D., Azizi, Z., & White, J. A. (2025). Bias recognition and mitigation strategies in artificial intelligence healthcare applications. *NPJ Digital Medicine*, *8*(1), Article 154. https://doi.org/10.1038/s41746-025-01503-7

Ho, V. (2025, February 26). *6 ways AI is making a difference in the world*. Microsoft. https://news.microsoft.com/source/features/ai/6-ways-ai-is-making-a-difference-in-the-world/

Holmes, K. (2018). *Mismatch: How inclusion shapes design*. The MIT Press. https://doi.org/10.7551/mitpress/11647.001.0001

Holtzblatt, K., & Beyer, H. (1997). *Contextual design: Defining customer-centered systems*. Morgan Kaufmann. https://shop.elsevier.com/books/contextual-design/holtzblatt/978-0-08-050304-2

Kemp, A. (2024). *As Xbox innovates again, is the gaming industry finally getting accessibility right?* The Drum. https://www.thedrum.com/news/xbox-innovates-again-the-gaming-industry-finally-getting-accessibility-right

Kerdvibulvech, C., Jiang, X. (2025). Generative AI in Human-Computer Interaction: Enhancing User Interaction, Emotional Recognition, and Ethical Considerations. In: Degen, H., Ntoa, S. (eds) Artificial Intelligence in HCI. HCII 2025. Lecture Notes in Computer Science, vol 15821. Springer, Cham. https://doi.org/10.1007/978-3-031-93418-6_5

Lalitha, A. R. (2025). *Beyond compliance: A user-autonomy framework for inclusive and customizable web accessibility*. arXiv. https://doi.org/10.48550/arXiv.2506.10324

Lee, M., & Kahng, M. (2025). Data-Model Co-Evolution: Growing Test Sets to Refine LLM Behavior (v1) [Preprint]. arXiv. https://doi.org/10.48550/arXiv.2510.12728

Liu, Y., Lou, Y., & Huang, S. (2024). Enhancing user experience through machine learning-based personalized recommendation systems: Behavior data-driven UI design. *Journal of Advanced Computer Systems*, 4(10), 25-37. https://doi.org/10.54254/2755-2721/2024.17905

Lu, M. Y., Chen, B., Williamson, D. F. K., Chen, R. J., Zhao, M., Chow, A. K., Ikemura, K., Kim, A., Pouli, D., Patel, A., Soliman, A., Chen, C., Ding, T., Wang, J. J., & Gerber, G. (2024). A multimodal generative AI copilot for human pathology. *Nature, 634*(8033), 466–473. https://doi.org/10.1038/s41586-024-07618-3

Ma, S. P., Liang, A. S., Shah, S. J., Smith, M., Jeong, Y., & Devon-Sand, A. (2025). Ambient artificial intelligence scribes: utilization and impact on documentation time. *Journal of the American Medical Informatics Association, 32*(2), 381–385. https://doi.org/10.1093/jamia/ocae304

Martin, A., MacDonald, R. L., Jiang, P.-P., Ladewig, M., Cattiau, J., Heywood, R., Cave, R., Tobin, J., Nelson, P. C., & Tomanek, K. (2025). Project Euphonia: advancing inclusive speech recognition through expanded data collection and evaluation. *Frontiers in Language Sciences.* https://doi.org/10.3389/flang.2025.1569448

Maslej, N., Fattorini, L., Perrault, R., Gil, Y., Parli, V., Kariuki, N., Capstick, E., Reuel, A., Brynjolfsson, E., Etchemendy, J., Ligett, K., Lyons, T., Manyika, J., Niebles, J. C., Shoham, Y., Wald, R., Walsh, T., Hamrah, A., Santarlasci, L., ... & Lotufo, J. B. (2025). *Artificial Intelligence Index Report 2025*. [Preprint]. arXiv. https://arxiv.org/abs/2504.07139

Microsoft. (n.d.). *Seeing AI: A Visual Assistant for the Blind*. Retrieved [November 7, 2025] from https://www.seeingai.com/

Microsoft Design. (n.d.). *Microsoft Inclusive Design*. Retrieved [November 9, 2025], from https://inclusive.microsoft.design/

*Micro*soft Garage. (n.d.). *Xbox Adaptive Controller*. Microsoft. Retrieved *[November 9, 2025], from* https://www.microsoft.com/en-us/garage/wall-of-fame/xbox-adaptive-controller/

Mitchell, M., Wu, S., Zaldivar, A., Barnes, P., Vasserman, L., Hutchinson, B., Spitzer, E., Raji, I. D., & Gebru, T. (2019). Model cards for model reporting. In *Proceedings of the 2019 Conference on Fairness, Accountability, and Transparency (FAT* '19)* (Article 52, pp. 318–326). ACM. https://doi.org/10.48550/arXiv.1810.03993

Ohu, F. C., Burrell, D. N., & Jones, L. A. (2025). Public health risk management, policy, and ethical imperatives in the use of AI tools for mental health therapy. PMC Public Health Risk Management, Policy, and Ethical Imperatives in the Use of AI Tools for Mental Health Therapy, 12, Article 12609870. https://doi.org/10.3390/healthcare13212721

Rahaman, S. U., Abdul, M. J., & Patchipulusu, S. (2023). AI-driven empathy in UX design: Enhancing personalization and user experience through predictive analytics. *International Journal of Computer Engineering and Technology, 14*(2), 255–268. https://www.researchgate.net/publication/384730424_AI-Driven_Empathy_in_UX_Design_Enhancing_Personalization_and_User_Experience_Through_Predictive_Analytics

Sanders, E. B. N., & Stappers, P. J. (2008). Co-creation and the new landscapes of design. *Co-Design, 4*(1), 5–18. https://doi.org/10.1080/15710880701875068

Shaikh, S. (2023). *Seeing AI App Launches on Android – Including new and updated features and new languages*. Microsoft Accessibility Blog. https://blogs.microsoft.com/accessibility/seeing-ai-app-launches-on-android-including-new-and-updated-features-and-new-languages/

Shen, H., Knearem, T., Ghosh, R., Liu, M. X., Monroy-Hernández, A., Wu, T., Yang, D., Huang, Y., Mitra, T., Li, Y., & Hearst, M. (2025). Bidirectional Human-AI Alignment: Emerging Challenges and Opportunities. In *Proceedings of the Extended Abstracts of the CHI Conference on Human Factors in Computing Systems (CHI EA '25)* (Article 857, pp. 1–6). Association for Computing Machinery. https://doi.org/10.1145/3706599.3716291

Shin, Y. (2025). Toward human-centered artificial intelligence for users' digital well-being: Systematic review, synthesis, and future directions. *JMIR Human Factors, 12,* Article e69533. https://doi.org/10.2196/69533

Shrivastava, J. (2025). Human vs AI in UX design. *International Journal of Engineering Research & Management Technology, 12*(2), 143–160. https://www.ijermt.org/publication/78/153.%20_Apr%202025%20(2Aug)%20ijermt.pdf

Stanford University Institute for Human-Centered Artificial Intelligence (HAI). (2025). *The 2025 AI Index Report*. Stanford University. https://hai.stanford.edu/ai-index

Strielkowski, W., Grebennikova, V., Lisovskiy, A., Rakhimova, G., & Vasileva, T. (2025). AI-driven adaptive learning for sustainable educational transformation. *Sustainable Development, 33*(2), 1921–1947. https://doi.org/10.1002/sd.3221

Swan, H. (2025). *Guide to the inclusive design principles*. TetraLogical. https://tetralogical.com/blog/2025/10/07/guide-to-the-inclusive-design-principles/

Thieme, A., Hanratty, M., Lyons, M., Palacios, J. E., Marques, R. F., Morrison, C., & Doherty, G. (2023). Designing human-centered AI for mental health: Developing clinically relevant applications for online CBT treatment. *ACM Transactions on Computer-Human Interaction, 30*(2), 1–50. https://www.microsoft.com/en-us/research/wp-content/uploads/2022/09/ToCHI_paper_final_08_09_2022_updated.pdf

Varshney, K. (2018). *Introducing AI Fairness 360*. IBM Research Blog. https://research.ibm.com/blog/ai-fairness-360

Velush, L. (2025). *The AI revolution: How Microsoft Digital (IT) is responding with an AI Center of Excellence*. Microsoft Inside Track. https://www.microsoft.com/insidetrack/blog/the-ai-revolution-how-microsoft-digital-it-is-responding-with-an-ai-center-of-excellence/

Web Accessibility Initiative (WAI). (2024). *Introduction to web accessibility*. In S. L. Henry (Ed.), W3C. https://www.w3.org/WAI/fundamentals/accessibility-intro/

Web Content Accessibility Guidelines Working Group. (2025). *Web Content Accessibility Guidelines (WCAG) 2.1* (W3C Recommendation). W3C. https://www.w3.org/TR/WCAG21/

Woebot Health. (2025). *Woebot Health.* https://woebothealth.com/

Xiang, A., Andrews, J. T. A., Bourke, R. L., Thong, W., LaChance, J. M., Georgievski, T., Modas, A., Rahmattalabbi, A., Ba, Y., Nagpal, S., Papakyriakopoulos, O., Zhao, D., Xue, J., Matthews, V., Gong, L., & Hoag, A. T. (2025). Fair human-centric image dataset for ethical AI benchmarking. *Nature.* https://doi.org/10.1038/s41586-025-09716-2

Zhang, J., Shu, Y., & Yu, H. (2023). Fairness in design: A framework for facilitating ethical artificial intelligence designs. *International Journal of Crowd Science, 7*(1), 32–39. https://doi.org/10.26599/IJCS.2022.9100033

Zhu, Q., & Luo, J. (2024). Toward artificial empathy for human-centered design. *Journal of Mechanical Design, 146*(6), 061401. https://doi.org/10.1115/1.4064161

Zia, T. (2024). *Empathetic AI: Transforming mental healthcare and beyond with emotional intelligence*. Unite.AI. https://www.unite.ai/empathetic-ai-transforming-mental-healthcare-and-beyond-with-emotional-intelligence/

22. Rapid AI Prototyping Without Technical Debt

Cieden. (2024, September 21). *AI for rapid prototyping: A practical guide with playbooks from product leaders.* https://cieden.com/ai-for-rapid-prototyping

Consortium for Information and Software Quality. (2022). *The cost of poor software quality in the US: A 2022 report.* https://www.it-cisq.org/the-cost-of-poor-quality-software-in-the-us-a-2022-report/

Ficus Technologies. (2024). *AI-powered code generation in 2024: Can developers trust it for production?* https://ficustechnologies.com/insights/blog/ai-powered-code-generation-in-2024-can-developers-trust-it-for-production/

GitClear. (2024). *Coding on copilot: 2023 data suggests downward pressure on code quality.* *https://www.gitclear.com/coding_on_copilot_data_shows_ais_downward_pressure_on_code_quality*

Google Cloud. (2024). *2024 State of DevOps report: The evolution of platform engineering*. DORA (DevOps Research and Assessment). https://cloud.google.com/devops/state-of-devops

Jassy, A. (2024, April 11). *CEO Andy Jassy's 2023 letter to shareholders* [Annual shareholder letter]. Amazon. https://www.aboutamazon.com/news/company-news/amazon-ceo-andy-jassy-2023-letter-to-shareholders

Kruchten, P., Nord, R. L., & Ozkaya, I. (2019). *Managing Technical Debt: Reducing Friction in Software Development*. Addison-Wesley Professional.

Matthews, C. (2024, January 7). *A guide to AI prototyping for product managers.* Lenny's Newsletter. https://www.lennysnewsletter.com/p/a-guide-to-ai-prototyping-for-product

Nielsen Norman Group. (2024). *AI prototyping in real design contexts: Good from afar, but far from good.* https://www.nngroup.com/articles/ai-prototyping/

Protiviti. (2023). *CFOs should lead the charge in addressing technical debt.* https://www.protiviti.com/us-en/insights/flash-report-issue-75

23. The Agentic AI Systems

For deeper architectural insights and implementation patterns, I invite you to explore my ongoing work at: *substack.com/@lakshmiveeramani/posts*. For examples of agentic AI systems in different business sectors, see the chapter resources on the book website.

Anthropic. (2026). *Agent Skills - Claude API Docs.* Last accessed 2026-02-07. https://platform.claude.com/docs/en/agents-and-tools/agent-skills/overview*)*

LangChain. (2026). *Multi-agent - Docs.* Last accessed 2026-02-07. https://docs.langchain.com/oss/python/langchain/multi-agent

Veeramani, L. and Smiley, K. (2024, December 31). *AI Agents: Introduction. 6 'P's in AI Pods.* https://substack.com/@lakshmiveeramani/p-145854970

Veeramani, L. (2025, March 7). *What are NOT AI Agents?* 6 'P's in AI Pods. https://substack.com/@lakshmiveeramani/p-158414193

24. Stop Prompt Gambling: How to Govern AI-Human Collaboration

Rosenfeld, A. & Erk, K. (2018), Deep Neural Models of Semantic Shift

– END –

www.ingramcontent.com/pod-product-compliance
Lightning Source LLC
LaVergne TN
LVHW052219200926
848481LV00001B/154